1st EDITION

Perspectives on Modern World History

Building the Panama Canal

1st EDITION

Perspectives on Modern World History

Building the Panama Canal

Sylvia Engdahl

Editor

GREENHAVEN PRESS

A part of Gale, Cengage Learning

GALE CENGAGE Learning

Detroit • New York • San Francisco • New Haven, Conn • Waterville, Maine • London

GALE
CENGAGE Learning

Elizabeth Des Chenes, *Director, Publishing Solutions*

For more information, contact:
Greenhaven Press
27500 Drake Rd.
Farmington Hills, MI 48331-3535
Or you can visit our Internet site at gale.cengage.com.

Articles in Greenhaven Press anthologies are often edited for length to meet page requirements. In addition, original titles of these works are changed to clearly present the main thesis and to explicitly indicate the author's opinion. Every effort is made to ensure that Greenhaven Press accurately reflects the original intent of the authors. Every effort has been made to trace the owners of copyrighted material.

LIBRARY OF CONGRESS CATALOGING-IN-PUBLICATION DATA

Building the Panama Canal / Sylvia Engdahl, book editor.
p. cm. -- (Perspectives on modern world history)
Includes bibliographical references and index.
ISBN 978-0-7377-6364-5 (hardcover)
1. Panama Canal (Panama)--History. 2. Canals--Panama--Design and construction--History. 3. Canal Zone--History. I. Engdahl, Sylvia.
F1569.C2B84 2012
972.87'5--dc23 2012009231

Printed in the United States of America
1 2 3 4 5 6 7 16 15 14 13 12

CONTENTS

CHAPTER 2 Controversies Surrounding the Building of the Panama Canal

FOREWORD

> *"History cannot give us a program for the future, but it can give us a fuller understanding of ourselves, and of our common humanity, so that we can better face the future."*
>
> *—Robert Penn Warren, American poet and novelist*

The history of each nation is punctuated by momentous events that represent turning points for that nation, with an impact felt far beyond its borders. These events—displaying the full range of human capabilities, from violence, greed, and ignorance to heroism, courage, and strength—are nearly always complicated and multifaceted. Any student of history faces the challenge of grasping the many strands that constitute such world-changing events as wars, social movements, and environmental disasters. But understanding these significant historic events can be enhanced by exposure to a variety of perspectives, whether of people involved intimately or of ones observing from a distance of miles or years. Understanding can also be increased by learning about the controversies surrounding such events and exploring hot-button issues from multiple angles. Finally, true understanding of important historic events involves knowledge of the events' human impact—of the ways such events affected people in their everyday lives—all over the world.

Perspectives on Modern World History examines global historic events from the twentieth century onward by presenting analysis and observation from numerous vantage points. Each volume offers high school, early college level, and general interest readers a thematically arranged anthology of previously published materials that address a major historical event, with an emphasis on international coverage. Each volume opens with background information on the event, then presents the controversies surrounding that event, and concludes with first-person narratives from people who lived through the event or were affected by it. By providing primary sources from the time of the event, as well as relevant commentary surrounding the event, this series can be used to inform debate, help develop critical thinking skills, increase global awareness, and enhance an understanding of international perspectives on history.

Material in each volume is selected from a diverse range of sources, including journals, magazines, newspapers, nonfiction books, personal narratives, speeches, congressional testimony, government documents, pamphlets, organization newsletters, and position papers. Articles taken from these sources are carefully edited and introduced to provide context and background. Each volume of Perspectives on Modern World History includes an array of views on events of global significance. Much of the material comes from international sources and from US sources that provide extensive international coverage.

Each volume in the Perspectives on Modern World History series also includes:

- A full-color **world map**, offering context and geographic perspective.
- An annotated **table of contents** that provides a brief summary of each essay in the volume.
- An **introduction** specific to the volume topic.
- For each viewpoint, a brief **introduction** that has notes about the author and source of the viewpoint, and that provides a summary of its main points.
- Full-color **charts**, **graphs**, **maps**, and other visual representations.

- Informational **sidebars** that explore the lives of key individuals, give background on historical events, or explain scientific or technical concepts.
- A **glossary** that defines key terms, as needed.
- A **chronology** of important dates preceding, during, and immediately following the event.
- A **bibliography** of additional books, periodicals, and websites for further research.
- A comprehensive **subject index** that offers access to people, places, and events cited in the text.

Perspectives on Modern World History is designed for a broad spectrum of readers who want to learn more about not only history but also current events, political science, government, international relations, and sociology—students doing research for class assignments or debates, teachers and faculty seeking to supplement course materials, and others wanting to improve their understanding of history. Each volume of Perspectives on Modern World History is designed to illuminate a complicated event, to spark debate, and to show the human perspective behind the world's most significant happenings of recent decades.

INTRODUCTION

Nearly a hundred years ago, in the summer of 1914, the world changed. Two major events altered the relationship between nations and, ultimately, the lives of their people in a way that is all but forgotten except by the few individuals old enough to remember how things were before. The first of these events, on July 28, was the start of World War I, which had a far-reaching impact on European and American attitudes and customs. Little more than two weeks later, on August 14, came a long-anticipated occasion that had been widely expected not only to increase prosperity in nations throughout the world but, ironically, to establish lasting peace by bringing them closer together and facilitating trade between them. This milestone was the opening of the Panama Canal. David McCullough, in his 1977 book *The Path Between the Seas*, wrote:

> The creation of the Panama Canal was far more than a vast, unprecedented feat of engineering. It was a profoundly important historic event and a sweeping human drama not unlike that of war. Apart from wars, it represented the largest, most costly single effort ever before mounted anywhere on earth. . . . It affected the lives of tens of thousands of people at every level of society and of virtually every race and nationality. . . . It was both the crowning constructive effort . . . of the Victorian Era and the first grandiose and assertive show of American power at the dawn of the new century. And yet the passage of the first ship through the canal in the summer of 1914—the first voyage through the American land mass—marked the resolution of a dream as old as the voyages of [Christopher] Columbus.

Before the canal was built, in order to get from the Atlantic Ocean to the Pacific Ocean or vice versa every ship had to travel all the way around South America on a long and dangerous voyage. A passage across the Isthmus of Panama between North and South America would cut thousands of miles off the journey. The early explorers of the Americas had realized this, and various proposals for a canal had been made; in the 1890s a French attempt to build one had failed. Not until the United States undertook the project in 1903 was there a real prospect of bringing the dream to fulfillment.

It is hard now to grasp the magnitude of people's enthusiasm for the canal's construction and the hopes associated with its completion. To governments and commercial interests everywhere, it promised a revolution in international trade, as the cost of transporting goods from port to port would be dramatically decreased and all nations would benefit. To the US public, it meant far more. In the first place, it created pride—in a technological achievement at the very edge of the era's capabilities, and in being able to provide a great benefit to the world at large, a contribution no other nation was then in a position to accomplish. Beyond pride, the canal came to symbolize the ideals embodied in the work being done in Panama and the society believed to be developing among the US citizens there. However, the media that popularized the subject published only what audiences wanted to read; the canal zone society was actually less admirable than reported.

In his autobiography, former US president Theodore Roosevelt wrote:

> A finer body of men has never been gathered by any nation than the men who have done the work of building the Panama Canal; the conditions under which they have lived and have done their work have been better than in any similar work ever undertaken in the tropics;

> they have all felt an eager pride in their work; and they have made not only America but the whole world their debtors by what they have accomplished.

This was true, but it left out a good deal that today is recognized as significant. The men to whom Roosevelt and other writers referred were the white Americans, who were provided with far more benefits than they would have received in the United States at that time. The work was hard and dangerous, but safety was not then considered as important as it is now, and rights of workers now taken for granted were not yet established anywhere. However, the working conditions of the black employees—mostly West Indians—were deplorable, and they were given little opportunity to take pride in their work. The world is indeed indebted to all whose labor contributed to the canal's existence.

Because of the racism that prevailed, and because of controversy over the issue of whether Panama was adequately compensated for the US canal rights, some historians question whether the building of the canal should be viewed in as positive a light as it has traditionally been seen. But the facts remain that the canal was vital to the world and that a hundred years ago no country other than the United States could have built it. Whatever wrongs were involved—and there is disagreement among commentators as to the extent of them—the creation of the canal was of inestimable international value. Attitudes unacceptable today should be judged in the context of an era in which they were commonplace.

Some of the controversy existed right from the beginning. In 1903, when the United States was ready to build the canal after lengthy debates in Congress, the Isthmus of Panama belonged to Colombia. Colombia favored the project and a treaty had been negotiated, but at the last minute they refused to ratify it. Panamanian leaders wanted the canal to be built, and they wanted the money the United States was willing to pay for the right to build it. So they declared independence from Colombia through a brief, bloodless revolution—bloodless because President Roosevelt sent US gunboats to prevent Colombian troops from landing in Panama. Colombia was outraged, and many people, even in the United States, believed that the president had misused his authority. He was accused of instigating the revolution, but he denied this, pointing out that the Panamanians had attempted to revolt many times in the past.

"It seems to be the part of statesmanship in this dilemma to talk loudly about the benefits we confer upon the world's commerce by digging the canal and to regard our acquisition of the canal a closed incident," declared the *Saturday Evening Post* in a 1912 editorial. The same year, in his book *The Americans in Panama,* William R. Scott wrote, "It is hypocritical to profess that we made adequate compensation when we paid Panama for the Canal Zone. We must applaud President Roosevelt for taking the Canal Zone, but the failure to make reparation to Colombia is a conspicuous piece of self-deception and moral obliquity." Eventually, in 1921, the United States paid Colombia $25 million; but the nations of Latin America continued to resent the United States' role in Panama.

The United States paid the new Republic of Panama for the right to build, including the right to sovereignty over a Canal Zone extending five miles on each side of the canal itself. But much of the money went to the French, who had previously abandoned the area. Later it was questioned whether the annual lease payments to Panama were large enough. In 1977 journalist Marjory Burns Shanard wrote:

> The US has good title to the land originally held by the Panama Railroad. This land we bought from the French. . . . Another approximately one third . . . is land for which we paid damages to private owners. The

> remainder of the Canal Zone is that which we gained the right to use in perpetuity under the 1903 treaty. However, at this point in time nobody knows exactly which land is which.

Although the Panamanians were pleased by the prosperity the canal brought to them, as time went on they became more and more unhappy about having a strip controlled by the United States running through the middle of their country. They felt that the canal should be turned over to Panama. Some Americans agreed, but the majority believed that since the United States had paid the cost of building the canal there was no justification for giving it away. Many nations supported Panama's position, and it was evident that sooner or later there would be trouble. Far from promoting peace as originally expected, the canal had become the focus of conflict. So in 1977 another controversial treaty was signed, calling for a gradual withdrawal of the United States to be completed by the end of the century. On the last day of 1999, the canal became the property of Panama, an outcome that many people in the United States still view with dismay.

The Panama Canal was never intended to benefit the United States alone. What was written during its construction invariably centered on what it would mean to the world as a whole. That outlook was rare at the time, when travel was slow and difficult and all communication between continents other than cable messages had to be carried by ship. Much of the excitement the canal engendered arose from its stimulation of that outlook.

Complete electronic texts of dozens of books written while the canal was being built are available on the Internet, some of which are listed in this book's bibliography. Many writers extolled the canal during those years, and their wording may appear somewhat exaggerated to the modern eye. Hyperbole in expressing convictions was customary, but the convictions were sincere. "America has made many contributions to mankind; it has given the world many great and high gifts, but the Panama Canal will be the greatest contribution that any nation has ever made to the general welfare of mankind," said James A. Green in 1907, addressing a group of businessmen who had just visited the Canal Zone. And in a 1914 book, Willis J. Abbot spoke for most of his contemporaries when he wrote:

> If it be properly conducted in its completed state, managed and directed upon the broad principle that though paid for wholly by the United States it is to exist for the general good of all mankind, it should be, in the ages to come, the greatest glory attached to the American flag. . . . In tearing away the most difficult barrier that nature has placed in the way of world-wide trade, acquaintance, friendship, and peace, we have done a service to the cause of universal progress and civilization the worth of which the passage of time will never dim.

Greenland (Denmark)
Alaska (U.S.)
CANADA
UNITED STATES
ATLANTIC OCEAN
Azores Islands (Portugal)
MEXICO
Hawaii (U.S.)
Tropic of Cancer
see inset below
CAPE VERDE
PACIFIC OCEAN
French Guiana (France)
Galápagos Islands (Ecuador)
COLOMBIA
Equator
Line Islands (U.S.)
SURINAME
ECUADOR
American Samoa (U.S.)
PERU
BRAZIL
SAMOA
BOLIVIA
French Polynesia (France)
TONGA
Tropic of Capricorn
PARAGUAY
Pitcairn Islands (U.K.)
CHILE
URUGUAY
ARGENTINA
Falkland Islands (U.K.)
SOUTHERN OCEAN
Antarctic Circle
Gulf of Mexico
UNITED STATES
N
BAHAMAS
Tropic of Cancer
Turks and Caicos Islands (U.K.)
CUBA
U.S. Virgin Islands (U.S.)
British Virgin Islands (U.K.)
Anguilla (U.K.)
St. Martin (St. Maarten) (France & Neth. Antilles)
ANTIGUA AND BARBUDA
MEXICO
Cayman Islands (U.K.)
HAITI
BELIZE
JAMAICA
DOMINICAN REPUBLIC
Puerto Rico (U.S.)
ST. KITTS AND NEVIS
Montserrat (U.K.)
Guadeloupe (France)
DOMINICA
GUATEMALA
Caribbean Sea
Martinique (France)
HONDURAS
ST. LUCIA
BARBADOS
EL SALVADOR
ST. VINCENT AND THE GRENADINES
NICARAGUA
0 200 400 mi.
0 200 400 km
Aruba (Neth.)
Netherlands Antilles (Neth.)
GRENADA
TRINIDAD AND TOBAGO
COSTA RICA
Lake Maracaibo
PACIFIC OCEAN
PANAMA
VENEZUELA
GUYANA
COLOMBIA

RUSSIA
KAZAKHSTAN
MONGOLIA
GEORGIA
AZER.
ARMENIA
UZBEK.
TURKMEN.
KYRGYZSTAN
TAJIKISTAN
CHINA
NORTH KOREA
SOUTH KOREA
JAPAN
PACIFIC OCEAN
ISRAEL
IRAQ
IRAN
AFGHAN.
JORDAN
BAHRAIN
QATAR
KUWAIT
PAKISTAN
NEPAL
BHUTAN
ALGERIA
LIBYA
EGYPT
SAUDI ARABIA
U.A.E.
OMAN
INDIA
MYANMAR (BURMA)
BANGLADESH
LAOS
TAIWAN (Claimed by China)
CHAD
SUDAN
ERITREA
YEMEN
DJIBOUTI
THAILAND
VIETNAM
CAMBODIA
PHILIPPINES
MARSHALL ISLANDS
ETHIOPIA
CEN. AFR. REP.
CAMEROON
SRI LANKA
BRUNEI
MALAYSIA
FEDERATED STATES OF MICRONESIA
KIRIBATI
SOMALIA
MALDIVES
DEM. REP. OF THE CONGO
UGANDA
KENYA
GABON
CONGO
RWANDA
BURUNDI
TANZANIA
SÃO TOMÉ PRÍNCIPE
SINGAPORE
INDONESIA
NAURU
PAPUA NEW GUINEA
SOLOMON ISLANDS
TIMOR-LESTE
TUVALU
SEYCHELLES
INDIAN OCEAN
ANGOLA
MALAWI
COMOROS
ZAMBIA
MOZAMBIQUE
MADAGASCAR
MAURITIUS
Réunion (France)
ZIMBABWE
BOTSWANA
NAMIBIA
VANUATU
FIJI ISLANDS
New Caledonia (France)
AUSTRALIA
SOUTH AFRICA
SWAZILAND
LESOTHO
0 1000 2000 mi.
0 1000 2000 km
NEW ZEALAND
SOUTHERN OCEAN
ANTARCTICA
see inset below
Arctic Circle
Western Sahara (Morocco)
ALGERIA
MAURITANIA
MALI
NIGER
SENEGAL
GAMBIA
GUINEA-BISSAU
GUINEA
BURKINA FASO
BENIN
SIERRA LEONE
TOGO
CÔTE D'IVOIRE
GHANA
NIGERIA
LIBERIA
Gulf of Guinea
EQUATORIAL GUINEA
200 400 mi.
200 400 km
SWEDEN
FINLAND
NORWAY
0 200 400 mi.
0 200 400 km
ESTONIA
RUSSIA
North Sea
DENMARK
LATVIA
IRELAND
UNITED KINGDOM
LITHUANIA
RUSSIA
Baltic Sea
NETHERLANDS
BELARUS
POLAND
ATLANTIC OCEAN
GERMANY
BELGIUM
LUX.
CZECH REP.
UKRAINE
LIECH.
SLOVAKIA
MOLDOVA
Bay of Biscay
FRANCE
SWITZ.
AUSTRIA
HUNGARY
ROMANIA
SLOVENIA
CROATIA
BOS. AND HERZ.
SERBIA
ANDORRA
MONACO
ITALY
PORTUGAL
SAN MARINO
MONT.
BULGARIA
Black Sea
SPAIN
Corsica (France)
Sardinia (Italy)
VATICAN CITY
ALBANIA
MAC.
Gibraltar (U.K.)
Ceuta (Spain)
Melilla (Spain)
Balearic Isands (Spain)
Sicily (Italy)
GREECE
TURKEY
MOROCCO
ALGERIA
TUNISIA
MALTA
Mediterranean Sea
CYPRUS
SYRIA
Crete (Greece)
LEBANON

CHAPTER 1

Historical Background on the Building of the Panama Canal

VIEWPOINT 1

Overview of Building the Panama Canal

Stephanie Watson

The following selection explains that there had long been a need for a canal across Central America so that ships traveling between Europe and the Far East, or between the Atlantic and Pacific coasts of the United States, would not have to go all the way around South America. In 1903 the United States settled on a route for the canal through Panama, which was then a part of Colombia, and supported a revolution in Panama in order to gain the right to build it. It was necessary to eradicate the mosquitoes that carried fatal diseases before construction could begin. Workers and engineers also had to combat landslides while excavating many tons of rock over a period of ten years. But in 1914 the canal—hailed as a marvel of engineering—was completed. Because of its impact on international trade throughout the world, US president Theodore Roosevelt considered the building of the canal the most important action he had taken in foreign affairs. Stephanie Watson is an independent scholar and writer.

Photo on previous page: The Miraflores Locks are one of three lock systems on the Panama Canal. (© **Will & Deni McIntyre/Getty Images.**)

SOURCE. Stephanie Watson, *Science and Its Times*. Belmont, CA: The Gale Group, 2000, pp. 535–538.

In the early years of oceanic commerce, ships carrying goods between Europe and the Far East had to travel a long, circuitous 12,000-mile (19,308 km) route around the continent of South America. As early as the 1500s, Spanish rulers first explored the idea of creating a canal through the Isthmus of Panama to drastically reduce travel time. In 1903, a treaty between Panama and the United States finally paved the way for the construction of the Panama Canal, a massive feat of engineering which not only united two oceans, the Atlantic and the Pacific, but opened an invaluable artery for international trade.

While the construction of the Panama Canal was not undertaken until the early twentieth century, it was first envisioned several centuries earlier. Christopher Columbus (1451–1506) searched in vain for a passage between the North American and South American continents. In 1534, Charles I of Spain recognized the value of cutting a route through Panama in order to gain greater accessibility to the riches of Peru, Ecuador, and Asia. He ordered a survey of a proposed canal route through the Isthmus of Panama drawn up, but unrest in Europe put the project on permanent hold.

The Right to Build a Canal Through Panama

The first construction did not get underway for another three centuries, when a Frenchman named Ferdinand Marie de Lesseps (1805–1894) decided that his country should take responsibility for building an interoceanic canal. Lesseps, who had completed Egypt's Suez Canal just 10 years earlier, felt this new canal would be equally successful. His crew began work on a sea-level canal in 1882, but their equipment proved insufficient to cut through the rocky terrain, money was growing short, and diseases such as yellow fever and malaria decimated the French work force. Lesseps's company was left bankrupt, and the operation was forced to shut down with the canal still incomplete. In 1894, the French created the New Panama Canal Company to finish the task, and set out to find a buyer.

Strategically, the canal took on new importance, as the U.S. empire stretched from the Caribbean across the Pacific.

Meanwhile, America too had been eyeing the isthmus. President Theodore Roosevelt, newly at the country's helm after the assassination of President McKinley, began building up America's military might and revitalizing its navy. Strategically, the canal took on new importance, as the U.S. empire stretched from the Caribbean across the Pacific. At the time, it took a full 67 days for the U.S. battleship, the *Oregon,* to travel from San Francisco to the Caribbean. Roosevelt was determined to prove America's military superiority and show that his country could bridge the distance between the Atlantic and Pacific oceans.

A few years earlier, the U.S. Congress had chartered the Maritime Canal Co., under the leadership of millionaire J.P. Morgan, to build a canal in either Nicaragua or Panama. Nicaragua was chosen, but a financial panic in the U.S. caused operations to shut down within five years. Now, a battle waged in Congress between completing the Nicaragua project or taking on construction of the Panama canal. In 1902, Iowa Senator William Hepburn introduced a bill to initiate construction of the Nicaraguan canal, to which Senator John Spooner of Wisconsin attached an amendment that literally overrode the bill, providing for a canal at Panama instead.

The Spooner Act gave President Roosevelt $40 million to purchase the New Panama Canal Company from the French, but there was one more obstacle to overcome. Panama [which was then part of Colombia] wanted to sell the land to America, but Colombia refused. Roosevelt predicted there would soon be a revolution in Panama,

The Panama Canal Was Expected to Foster World Peace

In these days when there is so much discussion of permanent peace among all the nations of the world, it is well to note the possible significance of the canal in relation to this world-wide peace movement. Closely associated with this thought, although it may have a strange sound at first, is the military or naval value of the canal. Remembering, however, that a distinctly military or naval advantage of one country or group of countries may be a powerful influence for peace, the meaning of the canal in this respect is easily appreciated. In its announced policy the United States is constructing the canal for the strategic use or military protection of the country as well as for commercial advantages. By being able to move its naval fleets and squadrons rapidly from the Atlantic and Gulf coasts to the Pacific coast, or vice versa, the United States gains an extraordinary advantage from the canal, which must reduce to a minimum the possibilities or probabilities of its being engaged in war by foreign countries. . . . But the influence of the canal for peace along other than military or naval lines is far more impressive.

As more commerce is developed between nations, as more travel goes back and forth between the peoples of different countries, as more intimate acquaintance grows between their representative men, as they become more dependent, one upon the other, in trade, friendship, and acquaintance, so much the less is the danger of war between them or of the development of differences which might lead to war. The canal will so directly develop the trade and travel between the commercial and political centers . . . that it cannot fail to be forever a mighty and tangible argument and influence for peace.

SOURCE. *John Barrett,* Panama Canal: What It Is, What It Means. *Washington, DC: Pan American Union, 1913. pp. 102–103.*

and he was right. A local revolt put a new government in Panama, which the U.S. firmly supported.

In 1903, the United States and Panama signed the Hay-Bunau-Varilla Treaty by which the United States guaranteed Panama's independence, and secured the right to construct an interoceanic canal through the Isthmus of Panama. America was given a perpetual lease on the 10-mile (16 km) wide canal zone.

Obstacles to Canal Construction

There were several major obstacles standing in the way of successfully completing the canal. It was, first of all, a major feat of engineering. The crews had to dig through the Continental Divide, a ridge of land separating two oppositely flowing bodies of water. The canal was to be the largest earth dam ever built at the time, with the most massive canal locks and gates ever constructed.

> "The canal was to be the largest earth dam ever built . . . with the most massive canal locks and gates ever constructed."

Sanitation was also a potential problem. The warm Panamanian climate was a breeding ground for mosquitoes, which carried malaria and yellow fever. Early French crews lost an estimated 10,000 to 20,000 workers to yellow fever outbreaks between 1882 and 1888. The American government vowed this would not happen to their own crews and, to this end, sent in doctor William Gorgas (1854–1920) to examine the area. In 1900, U.S. Army tropical disease expert Walter Reed (1851–1902) had proven that yellow fever was spread by the female mosquito. Gorgas vowed to remove the mosquito population from the canal zone, a task easier said than done. His first job was to treat all standing or slow-moving bodies of water with a combination of oil and insecticide. He covered windows with wire screens and sent health workers door-to-door to search for mosquitoes and their eggs. He fumigated houses and quarantined sick workers. By December 1905, yellow fever was virtually eliminated from the Canal Zone.

Before the actual construction of the canal could begin, the administrative infrastructure had to be put in place. Congress set up a commission to control the canal area, overseeing all transactions. Locals were hired to pave roads, repair the ailing French buildings, lay a new railroad track to accommodate American train cars, and put in a working water and sanitation system.

Building the Canal

Next, the commission hired its chief engineer, John F. Wallace, to oversee the project. Wallace began his excavation at Culebra Mountain. Culebra Cut was a 10-mile (16 km) stretch through the rockiest terrain along the canal route. Wallace quickly tired of the commission's layers of red tape, which were draining his time and patience, and petitioned Congress for a new governing body. Roosevelt quickly fired the commission, replacing them with seven new members. Wallace returned to the canal site, but ended up resigning a short time later.

In 1905, Wallace was replaced by John Stevens, and building of the canal began in earnest. Stevens first had to decide what type of canal to build—at sea level, as the French had begun, or by using locks—closed-off sections of the water used to raise and lower levels as ships moved through. He decided on a lock canal, a decision which was seconded by President Roosevelt.

But just when work seemed to be finally progressing, Stevens sent a letter to the president, indicating that he was not "anxious to continue in service." His resignation was accepted, and yet another replacement engineer was sought. Roosevelt appointed Army Lieutenant George Washington Goethals, whom he knew would exercise tight control and keep the project within the watchful eye of the U.S. Government. Quickly, Goethals organized his team, introducing a system to cut costs and keep everyone on track.

By 1907, more than 39,000 people were hard at work digging through the rock of the Culebra Cut. But there

The Miraflores Locks under construction in 1913. (© Roger Viollet/Getty Images.)

were problems. When the French had excavated Culebra Cut years earlier, their method involved chopping the tops off of hills and piling the dirt on either side, which led to mudslides. Indeed, the area was plagued by landslides. The American engineers were forced to increase the amount of rock and clay they excavated and adjust the angle of their cut in order to overcome the problem. In 1908, changes were made to the initial design of the canal because of unforeseen problems. The width was increased from 200 feet (61 m) to 300 feet (91 m) and the size of the locks was increased from 95 (29 m) to 110 feet (33 m).

Completion of the Canal

When completed, the canal stretched 50 miles (80 km), through three sets of locks. At its Atlantic entrance, a ship would pass through a 7-mile (11 km) dredged channel in Limón Bay. It would then travel 11.5 miles (18 km) to two parallel sets of locks at the town of Gatun, which were fueled by an enormous man-made lake. Each of the locks would raise or lower ships 85 feet (26 m). Then it was another 32 miles (51 km) through a channel in Gatun Lake to Camboa, where the Culebra Cut began. Another set of two locks located at Pedro Miguel and at Miraflores on the Pacific side would then lower the ship to sea level. Water flowed in and out of the locks through huge culverts, or drains, in the walls of the locks. A railroad-like locomotive served as a tow to pull the ships through the locks.

Ten years after construction had begun, the canal was finally completed in August of 1914 at a cost of approximately $387 million. The first ship to cross through was the concrete vessel *Cristobal*, however the first publicized voyage was that of the freighter *Ancon*. Unfortunately, Europe and the United States had just entered into World War I, so traffic across the canal was initially light. At first, around 2,000 ships per year passed through but by the end of the war, that number rose to 5,000. Soon, the Panama Canal was accommodating nearly all of the world's interoceanic trade vessels.

> "The Panama Canal would not only prove a modern marvel of engineering, it was to open up a whole new world in international trade."

Transfer of the Canal to Panama

In 1977, the United States and Panama signed two treaties granting the United States control of the canal until the end of the twentieth century. The U.S. was to oversee management, operation, and defense for the canal, and American ships would be free to travel back and forth across it. On December 31, 1999, just as the new millennium was about to dawn, the U.S. handed control of the Canal back over to Panama. At the ceremony, President

Jimmy Carter, who oversaw the initial transfer treaty in 1977, told Panama's President Mireya Moscoso, "It's yours."

The Panama Canal would not only prove a modern marvel of engineering, it was to open up a whole new world in international trade. Roosevelt once said of his achievement, "The canal was by far the most important action I took in foreign affairs during the time I was President. When nobody could or would exercise efficient authority, I exercised it."

VIEWPOINT 2

Preparing for Construction of the Canal

Panama Canal Authority

The following excerpt provides a brief history of preparations made for the construction of the Panama Canal. There was controversy about the best location for the canal. Panama was finally chosen over Nicaragua, in part because the route was shorter, even though France had attempted to build a canal there and had failed. Panama belonged to Colombia but declared its independence, backed by US military forces. A treaty granting the United States the canal concession was then signed. The first obstacle to be overcome was disease, which had killed or sickened many workers. Making provisions for housing and feeding thousands of people and then recruiting a work force was also a large task. Ultimately, the construction went well, and the completed canal was viewed as a major accomplishment of world civilization. The Panama Canal Authority is the Panamanian organization that now operates the canal.

SOURCE. "A History of the Panama Canal," Pancanal.com, n.d.

Panama was enveloped in its own "miasmal mist" of failure following the French Canal adventure. The second Walker Commission, the U.S. Isthmian Canal Commission of 1899–1902, ordered by President [William] McKinley, favored a Nicaragua route, as did both popular and official U.S. support. Panama seemed clothed in defeat, while Nicaragua was regarded as a clean slate for an all-American canal project.

Following President McKinley's assassination [in 1901], Theodore Roosevelt became president. For him, there was no romance about the project, none of this nonsense about following a dream. The canal was practical, vital and indispensable to the U.S. destiny as a global power with supremacy over both its coastal oceans. . . .

The canal was practical, vital and indispensable to the U.S. destiny as a global power with supremacy over both its coastal oceans.

Compagnie Nouvelle [a French company that owned assets in Panama] held a stockholders meeting in Paris, and, fearing to get left out in the cold with their proposed deal with the Americans, ascribed a new value to their Panama assets of $40,000,000. This just happened to be the value put upon them by the Americans. . . . The House [US House of Representatives], however, passed the Hepburn Bill favoring Nicaragua—two votes short of unanimous.

Through this, the White House had maintained silence; however, following the House vote, Roosevelt . . . let it be known that he wanted the French offer accepted and that the Commission was to provide a supplementary report unanimously favoring the Panama route. . . .

President Roosevelt submitted the supplementary report to Congress in January 1902. Wisconsin Senator John Coit Spooner introduced an amendment to the Hepburn Bill authorizing the president to acquire the French company's assets and concessions for a maximum price of $40,000,000. The bill stated that if arrangements could not be agreed upon between the United States and Colombia within "a reasonable time," the President would be authorized to seek an agreement for the alternate route through Nicaragua. . . .

Senator Mark Hanna was also in favor of the Panama route for technical reasons, reasons already provided in engineering reports. The Panama waterway would be shorter, straighter, take less time to transit, would require fewer locks, had better harbors, already had a railroad and would cost less to run.

Hanna's speech and support before the Senate were impressive, but not enough so to change the number of votes required. But it was [French engineer Philippe] Bunau-Varilla who turned the tide. To each senator he mailed a letter enclosing a one-centavo stamp showing a Nicaraguan landscape. In the background, the famous Momotombo volcano was depicted in full eruption. The stamp clearly pointed out the differences between the two countries—one with active volcanoes, the other comparatively stable. On June 19, 1902, the Senate vote favored a Panama canal route by just eight votes. . . .

The Panamanian Revolution

With the route decided, it was now time to begin negotiations with Colombia for a concession to build a canal through the Colombian province of Panama. The resulting Hay-Herran Treaty, developed by Colombian charge d'affaires Dr. Tomas Herran and U.S. Secretary of State John Hay, was rejected by Colombia. Roosevelt, reportedly furious, was not inclined to continue negotiations.

Impatient to build the canal, Roosevelt supported Panama's independence movement. And he was willing to put forth a show of military force, dispatching warships to both sides of the Isthmus . . . thus effectively blocking the sea approaches. Troops not only protected the railroad, but were also sent into the interior to block access from those areas. A land approach by a Colombian force of 2,000 was defeated by the Darien jungle and forced to turn back.

Roosevelt would later boast that ". . . I took the isthmus, started the canal and then left Congress not to debate the canal, but to debate me." Without the U.S. military presence it is doubtful that the Panama independence movement would have succeeded.

Without the U.S. military presence it is doubtful that the Panama independence movement would have succeeded.

Panama declared independence from Colombia on November 3, 1903. The Hay-Bunau-Varilla Treaty was negotiated by the new republic's "Envoy Extraordinary and Minister Plenipotentiary" Philippe Bunau-Varilla with John Hay. The new treaty was sent to Panama for ratification. The treaty granted to the United States as if sovereign a canal concession in perpetuity to a canal zone 10 miles wide, 5 miles on either side of the Canal prism line. Whether they liked it or not, the founders of Panama had little choice but to accede, as to refuse would have withdrawn all U.S. support from the fledgling republic. . . .

Eliminating Disease

The beginning of the U.S. canal construction effort dates from May 4, 1904. . . . Chief Sanitary Officer Dr. William Crawford Gorgas and his staff were among the first to arrive and set up operations.

Medical researchers at around this time were becoming more receptive to the idea of a relationship between mosquitoes and malaria and yellow fever. . . . The great yellow fever discoveries in Cuba in 1900 were the work of Dr. Walter Reed, who happened at the time to be Gorgas's commanding officer, who proved that *Stegomyia fasciata* [a mosquito species] was the carrier, debunking all previous theories, including the belief

Chief Sanitary Officer Dr. William Crawford Gorgas was responsible for eradicating yellow fever in the Panama Canal area. (© **Bettmann/Corbis/AP Images.**)

that "fomites," the term used for the soiled clothes or bedding of yellow fever victims, could spread the disease. Gorgas, himself a yellow fever survivor and thus immune to the disease, was a particularly valuable member of the medical team. Still skeptical, however, he suggested to Reed that, to prove the theory once and for all, Havana needed to be rid of the *Stegomyia fasciata*. . . . The eradication procedures didn't just kill off the *Stegomyia fasciata*, but reduced the *Anopheles* [a mosquito genus] population as well, thus decreasing malaria cases by more than half. These same techniques were what Gorgas brought with him to Panama in 1904. . . .

For Gorgas, it was urgent to get a jump on mosquito eradication before new, non-immune workers arrived and became infected. . . .

The work to combat yellow fever included screening windows and doors, house-by-house fumigation of Panama City and Colon [a Panama port city] and weekly oiling of cisterns and cesspools. A most important advance was providing running water to Panama City, Colon and other townsites to do away with the need for the domestic water containers that served as perfect breeding sites for the yellow fever vector mosquito.

As a result of Gorgas's crusade, yellow fever was completely and permanently wiped out on the Isthmus, with the last case reported in Panama City on November 11, 1905.

Malaria, unlike yellow fever, does not confer immunity. With the disease endemic on the Isthmus, there were repeated opportunities to lay its victims low by debilitation or death. It actually was the cause of more deaths during the French and U.S. construction periods than was yellow fever. During the first year of the American effort, 1905, nearly all of the American force, including Gorgas, had contracted malaria after only a month on the Isthmus. Gorgas was to say, "If we can control malaria, I feel very little anxiety about other diseases. If we do not control malaria our mortality is going to be heavy." . . .

> Reducing and eradicating the swarms of malarial mosquitoes was a huge task.

Research, however, revealed valuable information. Knowing that the *Anopheles* mosquito cannot fly far without lighting on some sort of vegetation, 200-yard-wide areas were cleared around where people lived and worked. Sanitation teams drained more than 100 square miles of swamp, built nearly a thousand miles of earthen ditching, some 300 miles of concrete ditch, 200 miles of rock-filled trench, almost 200 miles of tile drain, cut hundreds of acres of wild vegetation, sprayed standing water with thousands of gallons of oil, hatched and released thousands of minnows to eat the

Anopheles larvae and bred spiders, ants, lizards to feed on adult insects. . . . While these efforts covered only a small fraction of the Zone area, they efficiently reduced malaria incidence in populated areas. Two hundred and eleven employees died of malaria during fiscal year 1906–1907, declining significantly from a peak of 7.45 per 1,000 in 1906 to .30 per 1,000 in 1913. This achievement greatly increased American chances of canal-building success. . . .

Native villages and towns in the Canal Zone . . . were required to move. Legal owners thus required to vacate were compensated for their property. Many inhabitants were required to relocate with the filling of Gatun Lake. Many of these sites dated from early days of Chagres River navigation, when the route was a much used commercial route across the Isthmus . . .

Making Preparations

As Panama was, to say the least, insufficiently developed or equipped to support the additional population created by the growing Canal labor force, a great deal of planning went into providing proper housing and an adequate food supply. Virtually everything that was needed for Canal construction, from equipment and building supplies to a labor force and food, would have to be brought to the Isthmus and distributed efficiently along the line of the canal. The Panama Railroad, which [Chief Engineer John] Stevens saw at once to be the lifeline of Canal construction, was completely overhauled. The lightweight, inadequate and mismatched equipment of the French was replaced with the best and toughest available, for this railroad would not only distribute workers, materials and supplies, but also would haul away the dirt and rock excavated from the channel. . . . Heavier track, engines, freight cars, dump cars and refrigerator cars were ordered, and bridges, signals and sidings were upgraded and improved. Also required and recruited from the

Image on following page: This 1905 magazine cover shows the United States, represented by President Roosevelt and Uncle Sam, facing an initial obstacle in the creation of the Panama Canal: yellow fever. **("Uncle Sam Fights Yellow Jack." Courtesy of HarpWeek.)**

HARPER'S WEEKLY

A JOURNAL OF CIVILIZATION

VOL. XLIX. New York, Saturday, July 22, 1905 No. 2535

THE FIRST MOUNTAIN TO BE REMOVED

United States was a phalanx of trained engineers, switchmen, operators, mechanics, yard masters, train masters, dispatchers, superintendents and conductors to first put together the railroad, as all components were shipped "knocked down," and then operate it. . . .

The size of the labor force was tripled in six months under Stevens and whole communities, including housing, mess halls, hospitals, hotels, schools churches, cold storage, clubhouses and laundries, were built to accommodate them. Streets were paved in Colon and Panama City and water and sewage systems installed. At one time, nearly half of the 24,000-man work force was employed at constructing buildings.

Stevens also developed the ingenious system of Canal excavation and disposal of rock and soil, called "spoil." He devised a complex but very workable and efficient system of railroad tracks at different levels within the Cut. Spoil train schedules were coordinated to the level where the excavation work was being done. Spoil train capacity kept pace with the excavation work, keeping both trains and steam shovels efficiently employed at all times. . . .

> [President] Roosevelt . . . spoke of building the canal as a mighty battle involving both the national honor and that of the work force.

Now that the canal project was off the ground and going well, Roosevelt's feeling about it underwent an apparent change. While at first he viewed it as a political, commercial and military necessity, he could now afford to allow himself to be inspired by the "romance" of the situation engendered by its dramatic challenges of its structural design and the many assorted difficulties overcome. Roosevelt now spoke of building the canal as a mighty battle involving both the national honor and that of the work force. The first president to leave the continental United States while in office, he made a trip to Panama in November 1906 to see for himself how things were going. At the end of his last day there, he made an impromptu speech to several hundred Americans, including John Stevens. Excerpts from those remarks reveal his thinking at the time:

> . . . whoever you are, if you are doing your duty, the balance of the country is placed under obligation to you, just as it is to a soldier in a great war. The man who does his duty, no matter in what position he may be placed, is the man for the job. But to do your duty you must do a little more than just earn your salary. As I have looked at you and seen you work, seen what you have done and are doing, I have felt just exactly as I would feel to see the big men of our country carrying on a great war.
>
> This is one of the great works of the world. It is a greater work than you yourselves at the moment realized.
>
> In the Grand Army the spirit what appeals to me is the spirit of fellowship, of comradeship. If a man was a lieutenant general of the army or if he was the last recruit, the youngest recruit whose age would permit him to serve in the ranks, it makes no difference. If he did his duty well, he is a comrade, and recognized in every Grand Army post. And so it should be with you, whether you be chief engineer, superintendent, foreman, steam shovel man, machinist, clerk—this spirit of comradeship should prevail. . . .

Provisions for the Canal Workers

Recruitment of a labor force was a big problem at the beginning of canal construction. With Panama's relatively sparse population there was no surplus labor anywhere in the republic. It was recognized early on that all classes of labor would have to be recruited from outside and that most of the higher grades of skilled labor would have to be recruited from the United States. The average number of Americans working for the Canal during the construction period was a little over 5,000.

The islands of the Caribbean were the logical place to seek a labor force as the French had done some years before. However, when the French canal effort failed, many West Indian laborers, about 20,000 of them, were left stranded in Panama, to be repatriated at their governments' expense. This experience left both governments and laborers themselves loath to participate in the American effort. Authorities on the Island of Barbados finally authorized large-scale recruitment leading to a total recruitment of 19,900 laborers. . . . When news got out of the high wages and good living conditions on the Isthmus, there was no longer a need to recruit, and all agents were withdrawn in 1909. . . .

"Every effort was made to ensure adequate living standards, in accordance with standards of the time, for canal workers."

Providing food for more than 40,000 employees and their families in a country with little food production capability and few stores was a tremendous task at the beginning. With the goal in mind of maintaining a healthy and contented work force, the Isthmian Canal Commission imported food on the Panama Railroad steamers. They also started farms to grow fruits and vegetables, even plants and flowers, as well as farms to produce milk and eggs.

It was a difficult task in the beginning, but every effort was made to ensure adequate living standards, in accordance with standards of the time, for canal workers. Ice and cold storage warehouses were constructed, and a bakery and ice cream plant were set up. The Panama Railroad had refrigerated cars to provide distribution to settlements along the line of the canal.

Hotels or restaurants were established for the American bachelors. A number of mess halls were built for the

European laborers where meals were furnished at 40 cents per day. Kitchens were built for the West Indian laborers. Rations were furnished and cooked in these kitchens for 30 cents per day. . . .

The Opening of the Canal

The first complete Panama Canal passage by a self-propelled, oceangoing vessel took place on January 7, 1914. The *Alexandre La Valley*, an old French crane boat that had previously been brought from the Atlantic side now came through the Pacific locks. . . .

> Unlike any other such project on record, the American canal had cost less in dollars than estimated.

Plans were made for a grand celebration to appropriately mark the official opening of the Panama Canal on August 15, 1914. A fleet of international warships was to assemble off Hampton Roads on New Year's Day 1915, then sail to San Francisco through the Panama Canal, arriving in time for the opening of the Panama-Pacific International Exposition, a world's-fair type celebration. But no such grand opening occurred. Although the Panama Pacific Exposition went on as planned, World War I forced cancellation of planned festivities at the Canal. The grand opening was a modest affair with the Canal cement boat *Ancon*, piloted by Captain John A. Constantine, the Canal's first pilot, making the first official transit. There were no international dignitaries in attendance. . . .

Amazingly, unlike any other such project on record, the American canal had cost less in dollars than estimated, with the final figure some $23,000,000 below the 1907 estimate, in spite of landslides and a design change to a wider canal.

Even more amazing is that this huge, complex and unprecedented project was carried out without any of the scandal or corruption that often plagues such efforts, nor has any hint of scandal ever come to light in subsequent years. . . .

The following words of Theodore Roosevelt are engraved in a plaque on display in the Rotunda of the Administration Building [of the Panama Canal Authority], and more than anything else convey his personal philosophy and the spirit of his thinking about the achievement at Panama:

> It is not the critic who counts, not the man who points out how the strong man stumbled, or where the doer of deeds could have done them better. The credit belongs to the man who is actually in the arena; whose face is marred by dust and sweat and blood; who strives valiantly, who errs and comes short again and again; who knows the great enthusiasms, the great devotions, and spends himself in a worthy cause; who, at the best, knows in the end the triumph of high achievement; and who, at the worst, if he fails, at least fails while daring greatly, so that his place shall never be with those cold and timid souls who know neither victory nor defeat.

David McCullough, in his book *The Path Between the Seas,* wrote:

> The creation of a water passage across Panama was one of the supreme human achievements of all time, the culmination of a heroic dream of over four hundred years and of more than twenty years of phenomenal effort and sacrifice. The fifty miles between the oceans were among the hardest ever won by human effort and ingenuity, and no statistics on tonnage or tolls can begin to convey the grandeur of what was accomplished. Primarily the canal is an expression of that old and noble desire to bridge the divide, to bring people together. It is a work of civilization.

VIEWPOINT 3

Deciding on the Design of the Canal

Duncan E. McKinlay

The following excerpt describes the heated debate over the issue of whether to build a sea-level canal, which would have required excavating the ground to that level all the way across the Isthmus of Panama, or a lock canal, which meant building a dam across the Chagres River to create a higher-level lake and a series of locks leading to sea level. The lock design would be less expensive and twice as fast to build; moreover, because at some seasons the river rose too high and with too strong a current to flow safely into a canal, dams would be needed anyway, the author says. Because of improved machinery, the lock was completed ahead of schedule. The saving on shipping costs made possible by the shorter route between the oceans was expected to far exceed the canal's cost. Duncan McKinlay, who wrote this description before the canal was finished, was a congressman from California who visited Panama with the Interstate Committee of the US House of Representatives.

SOURCE. Duncan E. McKinlay, *The Panama Canal.* San Francisco: Whitaker & Ray-Wiggin, 1912, pp. 28–39. Courtesy of Whitaker & Ray-Wiggin.

While the work of sanitization [in the Canal Zone] was underway, the President of the United States [Theodore Roosevelt] was taking counsel with a board of engineers as to the type of canal that should be constructed. As usual in all such matters, the authorities were about equally divided, half of the engineers being strongly in favor of a sea-level canal, and the other half advocating what was called a lock canal.

The Two Types of Canal

The difference between the two types of canal is this: A sea-level canal contemplated an excavation from shore to shore at the level of the sea; a lock canal contemplated the construction of a great dam across the valley of the Chagres and the course of the Chagres river, which dam would have the effect of holding the waters of the Chagres river. The accumulation of those waters in time would form a lake, the surface of which lake, of course, would be considerably above the level of the sea on either side. The dam would necessarily have to be surmounted through the agency of locks.

After much controversy and bickering, and a great deal of muck-raking by the newspapers and magazines of the United States and Europe, the plan of a lock canal was finally adopted. This plan contemplated the impoundment of the waters of the Chagres river by a dam constructed at Gatun, a little village about three and one-half miles inland from the shore of Limon bay. This dam when finished would be 7,700 feet in length, half a mile in width at the base, and 135 feet in height. It was designed that this dam should hold the waters of the lake at a height of 85 feet above sea-level, but it was constructed 50 feet higher so that all danger might be obviated in case of excessive floods.

The plan of the canal contemplated that this dam should be surmounted by three locks constructed in pairs, so that in case one series of locks became impaired the other could be used, or ships might pass up one side and down the other at the same time. . . . When completed, this series of locks constructed of concrete would be more than 3000 feet in length and about 250 feet in width, without doubt the largest concrete formation ever constructed.

The engineers of the Panama Commission give four reasons for the adoption of the lock system instead of the sea-level type. In the first place, it would take twice as long to construct a sea-level canal as it would a lock canal. Secondly, it would cost twice as much money, and as the lock canal system is costing nearly four hundred millions of dollars, the difference in cost would be a great obstacle to the construction of the other type of canal. The third reason was that in case a sea-level canal was constructed it would be necessary to place locks somewhere along its course because of the fact of the variation of tides between the Atlantic and the Pacific Oceans. . . . Great variation would cause a current to rush through the course of the canal so great that locks would be required for its control.

> The safety of the lock system would be much greater than that of the sea-level type.

But the fourth was the most potent reason of all why the lock system was adopted. On the Isthmus of Panama the rainfall amounts to 130 to 150 inches annually. Sometimes the precipitation will amount to 10 or 12 inches in twenty-four hours. The Chagres river is the only agency for the drainage of a vast area of water-shed in the Caribbean sea. Therefore, at times the Chagres river might be a small, inconsequential stream that a boy could wade across, and yet before twenty-four hours had elapsed, because of a heavy rainfall, it might have swelled into a raging torrent that would wreck the strongest battleship of the American navy. The large volume of water discharged by the Chagres river could not be turned into the canal proper, as the currents and the rush of flood waters would soon impair the banks of the canal.

The Lock System Adopted

Therefore it would be necessary, under the sea-level type of canal, to construct a series of embankments and dams that would be far more expensive to build and keep in repair than would be one great dam over the course of the Chagres river. Besides, the safety of the lock system would be much greater than that of the sea-level type. These were the reasons which finally controlled the determination of the engineers to construct a lock system of canal.

After the type of canal was decided upon, the next step was the assemblage of the force of laborers and the mechanical appliances necessary for the physical operations. . . .

Great difficulties were encountered in perfecting the proper kind of an organization to successfully complete this stupendous project. The engineers taken from private life and entrusted with the work, after a little experience on the Isthmus, would be offered greater inducements to abandon their Governmental employment and take some other position, generally far more lucrative, in the United States. . . .

Army Engineers Installed

This tendency of the civil engineers to leave their employment caused much concern to the President and Congress, and finally President Roosevelt, with his characteristic acumen, decided that he would place the work of canal construction under the army engineers entirely. . . .

Experience has proved that President Roosevelt's judgment was correct, for the work has gone on since the reorganization of the commission with the regu-

PROFILE DIAGRAM OF THE PANAMA CANAL

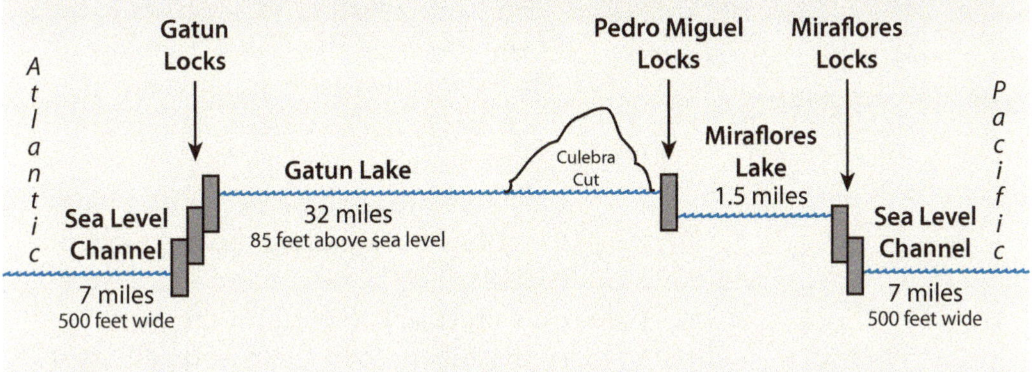

Taken from: William R. Scott, *The Americans in Panama*. New York: Statler, 1912. p. 156. www.openlibrary.org.

larity of a machine. There has hardly been a stop or a break at any point along the line of operations. Colonel G.W. Goethals, one of the most successful of the army engineers, was placed at the head of the Canal Commission and given full charge, and his work has been so successful that he has demonstrated his ability to command and to control the operations placed in his charge to the satisfaction of the great powers that gave him his commission.

His first step upon being placed in control was to provide the means of feeding and caring for an army of from 25,000 to 40,000 men. A bake shop was built at Crystobal [Cristobal, a port in the Atlantic side of the canal], out of which 30,000 loaves of bread are turned twice a day if necessary, and a batch of pies and cakes in proportion. Storage warehouses have been built for the storage of meats and vegetables and various other supplies that are brought from the north by shiploads. Ice plants have been constructed so that ice may be distributed up and down the line of operations. Every morning at 3 o'clock a supply train leaves Colon [a Panama port city], and furnishes every camp along the line of the

> The employees of the Canal Zone are as well supplied with rations and materials as they would be on an army reservation.

canal with fresh supplies for the day's consumption.

Thus, under army supervision the employees of the Canal Zone are as well supplied with rations and materials as they would be on an army reservation.

Following these necessary preparations for handling the big force of men, came the assemblage of the machinery and the mechanical implements necessary to perform the work. . . .

The stationary machinery is of the best quality that genius and money can construct, and so effective have been these means of labor saving that the work has been accelerated from time to time until it is now a realized fact that the canal will be actually constructed a year and a half ahead of time. . . .

The Gatun Dam

The Gatun dam is being constructed by hydraulic process through the instrumentality of suction pumps, which suck up the slime and the debris from the course of the Chagres river and the swamps and morass through which the canal is being constructed. This debris and this water are sucked up and allowed to run along the center of the dam, the water running off and the solid matter congealing there, and by this hydraulic process that great structure will be formed.

The traveler upon the Isthmus today, if standing upon an eminence overlooking the cut through Culebra hill, would imagine himself on a height overlooking an industrial city like Pittsburg. There are scenes of such immense activity on every side that he forgets he is in a remote part of the world far from his home, and that he is actually standing upon an eminence in the tropics.

The development of labor-saving machinery has been so marked since the construction of the canal was

The Gatun Dam controls the water level and provides electricity for the Panama Canal next to the Atlantic entrance. (© AP Images/Tomas van Houtryve.)

actually commenced that each month's work has marked an increase in the amount of debris excavated from the canal prism. When the [US] Government began operations in 1906, the engineers had before them the task of excavating 110,000,000 cubic yards. Their first month's operations were very successful, and they reported at the end of the month an excavation of about 250,000 cubic yards. They estimated that if they could keep up this amount of work through each month they could finish the canal at a certain time; but the carping yellow newspapers and magazines of the United States and Europe were extremely skeptical of the ability of the Canal Commission to continue to turn out 250,000 cubic yards per month. The critics foretold that when the rainy season came more debris would be carried into the canal prism by floods than could be taken out by machinery in the dry season. At times this criticism grew very irksome and disagreeable to the commissioners. However, they kept their temper, and continued improving their machinery, and month by month the output grew greatly. . . .

The Future of the Canal

When the Panama Canal is completed the commerce and trade of the world will be revolutionized. San Francisco will be brought nearly 9,000 miles closer to New York than it is today and European ports nearly 6,000 miles closer. It is estimated by statisticians skilled in transportation and in carrier service, that the cost of transporting the great mass of bulky products from the Pacific Coast to Eastern seabords of the United States and to European points will be reduced nearly two-thirds. . . .

> When the Panama Canal is completed the commerce and trade of the world will be revolutionized.

It is estimated that this saving of freight on timber alone, which is still standing in California, would pay the cost of the canal, great as it is, three times over. We can hardly estimate the effect that this shortening of water rates will have on all the countries fronting the Pacific Ocean. . . .

Men of even middle age today probably will live to see the fulfillment of the dreams and prophecies of the olden time in the opening up of our coasts and land to ship commerce with every country on the globe.

In ancient days it was the fact that seas divided nations, because of the difficulty of ocean travel. In those days the only safe routes were those over the land, but in this modern time of gigantic ocean vessels, capable of carrying thousands of passengers and hundreds of thousands of tons of freight, water travel and transportation is the cheapest and most agreeable of all forms. And therefore, today it is a fact that seas unite the countries of the world instead of dividing them.

Women Play an Important Role in the Canal Project

Wenona Marlin

In the following 1912 article, a writer relates the positive impact women were having in the Panama Canal Zone. The author explains that American workers were not willing to stay in Panama unless it was safe to bring their wives and children. Some American women went to Panama on their own to serve as nurses in the hospitals or as teachers, and many more came as brides sent for by workers who found it to be a suitable place for families. Both men and women were enthusiastic about the canal, she reported, and the women took pride in contributing to the project by making comfortable homes for their husbands. Wenona Marlin was the author of a book of short stories published in 1912 as well as articles such as this one for the *New York Times*.

SOURCE. Wenona Marlin, "Women in Making the Canal," *New York Times* archive, September 22, 1912.

We women have been the main thing in "making the dirt fly" on the Panama Canal. I know you are surprised to learn this. I was. . . .

After Uncle Sam had taken the Panama Canal off the hands of the French company and put the best available men in charge of the affairs of the canal, there was great consternation and dismay when one of the men holding a most important post gave up his position after a short stay on the Isthmus. Why? Well, some say it was because he did not wish to inflict the inconveniences of a tropical life upon the members of his family. It was a "tempest in a teapot" for a short time. It made people think—among these was Uncle Sam. The result of all this thinking might be summed up in the fact that it was decided that if the Panama Canal were to be built it would first have to be made a safe, comfortable and homelike place for men and women—not a place of exile and death.

> The Panama Canal . . . would have to be made a safe, comfortable and homelike place for men and women.

How the Canal Zone was "cleaned up" is also history, Col. Gorgas and his "housecleaning bee" became known the world over. It is said that Gorgas paid compliments to the ladies, played with the babies' toes—and turned them all into the streets until their houses were thoroughly disinfected.

Since then the women have aided in enforcing sanitary regulations. The men were careless frequently. They didn't care for themselves; they didn't fear anything. But when it came to the safety of their wives and children—a mosquito would cause a panic. The Canal Zone was changed from a deathtrap to a modern health resort. Women were there. Women were arriving on every steamer. . . .

Young women were sent by the United States Government to the Zone to find out what was most needed for the men and their families. They made their report. Many of their ideas have been adopted. Proper amusements and social life came to the front. Eight clubs were organized for the women and federated as in the United States, and since then much interest and enthusiasm has been shown in the club work. The Y.M.C.A. is really a clubhouse. There are to-day libraries, books, magazines, newspapers, schools, churches, clubs, hotels—all first-class in every way—and women. Instead of lawlessness there is peace; health instead of death; happy wives instead of scarlet women.

Influence of American Nurses

A word should be spoken (although a book could be written) of the work done by the nurses in the hospitals. Although the dread fever was extinct, stamped out, there were accidents among the 35,000 workmen and the death rate is about that in any city of an equal number of inhabitants—lower than some cities. The kind word and soft hand of the nurse has been a great aid to the "home-sick" boy down in Panama—possibly helped more than the medicine.

> "The kind word and soft hand of the nurse has been a great aid to the 'homesick' boy down in Panama."

A number of nurses went to Panama when sanitary regulations were almost unknown. The natives are immune from the diseases prevalent in tropical countries, and the foreigners were expected to get along the best way they could, and if they died—well, the funeral trains ran regularly to the Monkey Hill Cemetery, and it never lacked passengers. The nurses are a brave, devoted, untiring band of women. Just ask a canal man about the nurses and let him enthuse.

The public schools on the Canal Zone are the pride of all the Americans. "They can't be beat," said one enthusiastic father, for many years a canal employe.

Pretty, rosy-cheeked boys and girls attend the schools and they are happy and contented. It is a pleasure to recall their bright faces and quick replies to the questions of a stranger.

The teachers upon their arrival are assigned to one of the houses erected by the United States Government for bachelor girl employes and given a room (usually with a room-mate) for which she is entitled to select complete furnishings. The number of applicants is in excess of the vacancies. . . .

Brides and Homes

Many of the canal employees married after a short stay on the Isthmus. Uncle Sam stood ready with a home in his hand completely furnished—all but the wife. This helped to solve the question of keeping a permanent labor force on the canal, which was one of the most important problems that was ever laid before the Isthmian Commission.

Very often the "boy" sent for "the girl he left behind," and it has been no uncommon thing to meet a bride-to-be on the steamers plying between New York and Panama. On the trip down to the canal the Captain of the steamship said to me one day:

"Are you a bride-to-be?"

I expressed my astonishment at the question.

He continued. "You look suspicious. I have many brides on my steamer from Vermont, Iowa, and every place, going down to marry on the Zone."

"How romantic!" I exclaimed. "Just fancy, rushing over the country in a Pullman car, hurrying to the steamer traveling thousands of miles on the ocean, and going to a paradise—formerly a jungle—to be clasped in the arms of the one man in the whole world, and all the while guided and guarded by little Cupid! . . ."

The houses for the canal employes are all built alike, but of different sizes, some being for one family only

Americans Living in Panama Had Luxuries They Lacked in the United States

A bride starts out life [in Panama] on a basis that means a rude jolt to her when the canal is finished and she returns to the United States. Young couples who have been treading the easy path of high salary, free rent, free water, light, and fuel, cheaper food, clothes and furniture, elastic class distinctions and plentiful though not efficient servants, must ever look back upon their Canal Zone experience as the particularly bright point in their careers. The withering blasts of social competition, high cost of living, and salaries from one to two thirds lower in the United States, will make the easy-going, over-generous life at Panama seem the "temps de luxe" [time of luxury] in their lives.

SOURCE. *William R. Scott,* The Americans in Panama. *New York: Statler, 1912. p. 210. www.openlibrary.org.*

and others containing several families. They are set up on blocks of granite or stone and are so constructed that the air circulates freely all through the house, from the lower floor to the roof, and the houses are all thoroughly screened from insects. The furniture as well as the house are supplied free of charge by the Government, and it is of a good substantial quality.

Shopping and Marketing

Shopping and marketing for the women on the Zone are comparatively simple matters, with Uncle Sam to assist by supervising the Commissary Department and the stores, with entrances for the gold (white) employes

on one side and the silver (colored) employes on the other side. Moderate prices prevail and every delicacy of the season, as well as the more substantial eatables, are for sale, most of which are purchased in New York and Canada and shipped in steamers having departments specially devised for this purpose, and all is under the supervision of the Government.

The native labor is obtainable at a low cost and every house has several servants, which gives the American women ample time to pursue their favorite recreation. It may be an afternoon tea on the veranda, seated in easy wicker chairs 'mid tropical flowers and orchids galore, and a lawn filled with waving palms and cocoanut trees. Then, too, there are beautiful drives and horseback riding, tennis, fishing, boating, picknicking, and bathing in both the Atlantic and Pacific, as my lady may choose, are all pleasant ways of enjoying life in "the Land of the Cocoanut Tree." . . .

Photo on previous page: Colonel David du Bose Gaillard (standing) was the head of the department of dredging and excavating, in Panama. Wives of officials building the canal often joined their husbands in the Canal Zone to help create a comfortable home life. (© Corbis.)

Social Life

The weekly dances at the Tivoli Hotel (Ancon) are something to be remembered by all who attend. The ladies are dressed beautifully and the gentlemen are usually attired in cool-looking white suits. The hotel is perfect in all its appointments and is run by the Government. Cultured and traveled people from all over the world are coming and going, which makes it the social centre of the Canal Zone.

Naturally one wishes to know more of the lives of the native Panamanian society women who live a quiet secluded life quite apart from the American women. At Las Savannas (near the City of Panama) the wealthy Panamanian families live in beautiful homes amid the tropical plants. Their social life is formal and has little of the spontaneity so well known among the Americans. They are mostly descendants of old Spanish families and some to-day bear the names of men who came to the

Isthmus of Panama with [Christopher] Columbus and [Spanish explorer Vasco Núñez de] Balboa centuries ago. The women have black or dark brown hair, and beautiful dark eyes, and the olive complexion of the Spanish beauties. Like the women of Spain they are charming, but are void of the wholesome naturalness and vivacity of the American women. Strict etiquette prevails and the inevitable chaperone is prominent always. It is interesting to note, however, that with the advent of the American women and the growing practice of foreign education, the life of these Spanish girls has become somewhat freer and there is a perceptible change, the more independent ways of the American women being adopted in some families. . . .

> The transplanting of the American woman has been successful.

As a whole, the transplanting of the American woman has been successful. The "big ditch" absorbs the interest and takes precedence. There is no limit to the enthusiasms of both the men and women. After two years' work the men on the canal receive a medal. Their wives do not, but they deserve a medal, for they have done their part in making real homes so that men may "make good" and win the medals in—the greatest undertaking of modern times—the building of the Panama Canal.

VIEWPOINT 5

How Ships Will Navigate the Completed Canal

Joseph Bucklin Bishop and Robert E. Peary

The Panama Canal is actually a bridge of water formed by Gatun Lake, with locks on either side descending to sea level, explains the following viewpoint. This description of the canal's various parts, describing to the public how the locks work, was published in 1913, when most of the work was completed but the canal had not yet been opened. It tells what the passage through the canal on a ship will be like and what the benefits of the canal are expected to be. Joseph Bucklin Bishop was a newspaper editor, the Secretary of the Isthmian Canal Commission in Washington, DC, and the author of thirteen books. He co-wrote the book from which this excerpt was taken with Robert E. Peary, a US naval officer and explorer known for his expedition to the North Pole.

SOURCE. Joseph Bucklin Bishop and Robert E. Peary, *Uncle Sam's Panama Canal and World History*. New York: J. Wanamaker, 1913, pp. 9, 31–46, 58. Courtesy of J. Wanamaker.

The official seal of the Isthmian Canal Commission bears the motto: "The Land Divided. The World United." That is a terse and accurate statement of what the Panama Canal accomplishes. It divides a hemisphere and by opening a new and shorter ocean highway brings the nations of the earth into closer intercourse with one another.

> The canal . . . is in reality a great bridge of water, 87 feet above the level of the sea.

In supplying this highway the United States has conferred a benefit upon mankind which has few equals in human history, the full measure of which time alone can reveal.

Great wars change the map of the world by their victories, but few have made more radical and far reaching changes than will follow this momentous victory of peace—a victory for human welfare and progress. . . .

Layout of the Canal

The line of the canal across the Isthmus runs through low lying land from the Atlantic for a distance of about 30 miles when it strikes the foothills of the mountains of the Continental Divide. The foothills and mountains cover a distance of nearly 9 miles. After passing this obstruction the canal line runs through low lying land to the Pacific.

From the mountains and foothills of the Divide, there stretches away toward the Atlantic on either side of a broad valley a line of hills. These hills approach each other as they near the Atlantic till they are within a mile and a half of each other at Gatun. The valley lies between the mountains and the hills in the shape of a horse shoe with its opening toward the Atlantic.

Into this valley the Chagres River and several other streams pour their waters. The canal plan places a great dam at the opening of the horse shoe and impounds all these waters, making a huge reservoir covering 164 square miles. When full, the level of this lake is 87 feet above the mean level of the sea.

If nature had intended to make a mountain lake on this part of the Isthmus, like the lakes among many other mountains of the world, she would have brought the lines of encircling hills together at this point and the thing would have been done, and the work of the canal builders would have been greatly simplified. They would have been required only to cut a passage through the Divide.

What they had to do was to cut a passage through the Divide and transport the material removed thirty miles and make a dam out of it for the closing of the valley. . . .

Gatun Lake

The canal itself is unlike any other in the world. It is in reality a great bridge of water, 87 feet above the level of the sea, extending nearly the entire width of the Isthmus. Gatun Lake constitutes the larger part of it.

> In supplying this [ocean] highway the United States has conferred a benefit upon mankind which has few equals in human history.

To reach the surface of the lake, vessels must be lifted 87 feet and this is done by means of locks, the largest ever built. There are three pairs of these, on each side of the Isthmus, making 12 in all, and their combined lift is 87 feet.

On the Atlantic side the three pairs are in series, like a flight of steps, and vessels pass from one directly to another.

On the Pacific side there are two pairs at one point and one at another, with a small lake about a mile in length between them.

The water of the Gatun Lake is held in place by the Gatun Dam, which is about a mile and a half long, measured on its crest, a half mile wide at its base, and 105 feet high.

It contains 21,000,000 cubic yards of material, half mingled earth and rock brought from Culebra Cut, and half an impermeable mixture of sand and clay pumped into it by hydraulic dredges. This latter material forms the core of the dam.

The spillway of the dam is a concrete-lined passage 285 feet wide and 1,200 feet long which has been constructed through a hill of rock which stands nearly in the center of the dam.

At its entrance from the lake a concrete dam in the form of the arc of a circle has been erected, and on the top of this gates are placed for regulating the level of water in the lake.

The spillway and the lock gates are in reality portions of the dam in that they hold back the water to its maximum level of 87 feet.

When the water reaches a level above 87 feet the gates are opened and the surplus water runs off.

Connected with the spillway is a hydroelectric plant, operated by water from the lake and capable of generating sufficient electricity to operate all the locks and other machinery of the canal, light the canal and Canal Zone villages, and operate and light the Panama Railroad.

Culebra Cut

Culebra Cut, which is the name of the passage through the mountain range, is an arm of Gatun Lake. It is nearly 9 miles long, its channel has a bottom width of 300 feet, and in making it 100,000,000 cubic yards of earth and rock have been removed, or nearly double the original estimate.

It has a depth of 45 feet of water throughout, and its water level is that of Gatun Lake. It extends from Bas Obispo, at the southern extremity of Gatun Lake and the valley of the Chagres, to Pedro Miguel, where the first pair of locks on the Pacific side are placed.

The water bridge, therefore, reaches from the Gatun dam, through the Culebra Cut to Pedro Miguel, a distance of about 34 miles. The lock gates and small dams on either side correspond to the dam, spillway and gates at Gatun in holding the waters of Gatun Lake in place.

The width of the channel in the water bridge varies from 1,000 feet in the larger portion of the Gatun Lake to 300 feet in Culebra Cut. It is nowhere less than 300 feet. The depth of water in the channel varies from 87 feet in the lake to 45 feet in Culebra Cut.

> The canal does not run east and west, as is quite generally supposed, but from northwest to southeast.

There are approach channels in both oceans, each 500 feet wide and 45 feet deep, extending from deep water in the ocean to Gatun locks on the Atlantic side and to Miraflores locks on the Pacific side. . . .

The line of the canal does not run east and west, as is quite generally supposed, but from northwest to southeast. The Pacific entrance is 22½ miles east of that on the Atlantic.

The entire length of the canal from deep water in the Atlantic to deep water in the Pacific is about 50 miles—from shore-line to shore-line, about 40 miles.

The locks are the largest in the world, and are all of the same useable dimensions—1,000 feet long and 110 feet wide. They can accommodate easily the largest ships yet constructed. Their walls are of massive concrete, from 45 to 60 feet wide at the base and reaching to a height of 81 feet. They contain about 5,000,000 cubic yards of concrete.

The lock gates are steel structures seven feet thick, 65 feet long, and from 47 to 82 feet high. They weigh from 390 to 730 tons each. Ninety-two leaves are required for the entire canal, the total weighing 60,000 tons. . . .

Passage Through the Locks

The best way in which to get a correct conception of the canal as a whole is to accompany a vessel from ocean to ocean as it passes through it.

Starting four miles at sea in the Atlantic, a vessel enters the approach channel and passing by the city of Colon [a Panama port city] advances to the lower lock at Gatun, the water in which is on a level with that of the sea and 45 feet in depth.

Before entering the lock the vessel is tied up to a long wall or pier which extends into the channel between the two sets of locks for a distance of about 1,000 feet.

Here it is turned over to the canal operators, who take complete charge of it and are responsible for it during its passage through the locks. They place a representative of their own on the bridge and another in the engine room.

The worst accidents which have occurred to vessels in locks have been caused by an engineer mistaking a signal from the bridge and going ahead when he should have gone backward, thus ramming and breaking a lock gate.

This is not possible at Panama. No vessel is allowed to pass through the locks under its own power.

Vessels are taken through by electric locomotives running in cogged tracks on the approach and lock walls. . . .

Having passed [safely] the chain, the vessel enters the lower lock. The gates are closed behind it, and through openings in the floor, water enters, filling the lock chamber to the level of the water in the lock above.

When this level is reached, the inflow of water is stopped. The gates between the lower lock and the one above are then opened, and the vessel is towed into the second lock.

When it is inside, the gates behind it are closed, and water is again let in through the floor openings till the level in the second lock is raised to that of the third

A container ship enters the Miraflores Locks in the Panama Canal. (© Arnold H. Draping/Time & Life Pictures/Getty Images.)

above. When this is accomplished the gates in front are opened, and the vessel is towed forward into the third and last lock, the water level of which is that of the lake. It is then ready to proceed under its own power on its journey through Gatun Lake.

A Picturesque Journey

The channel through Gatun Lake varies in width from 1,000 feet to 500 feet, and is about 24 miles in length. A vessel can steam at full speed through this till it reaches the entrance of Culebra Cut.

Here the Channel narrows to 300 feet at the bottom. The vessel still proceeds under its own power, but at a reduced speed through Culebra Cut.

The irregular lines of thickly clad hills which encircle the lake make the passage through it one of much natural

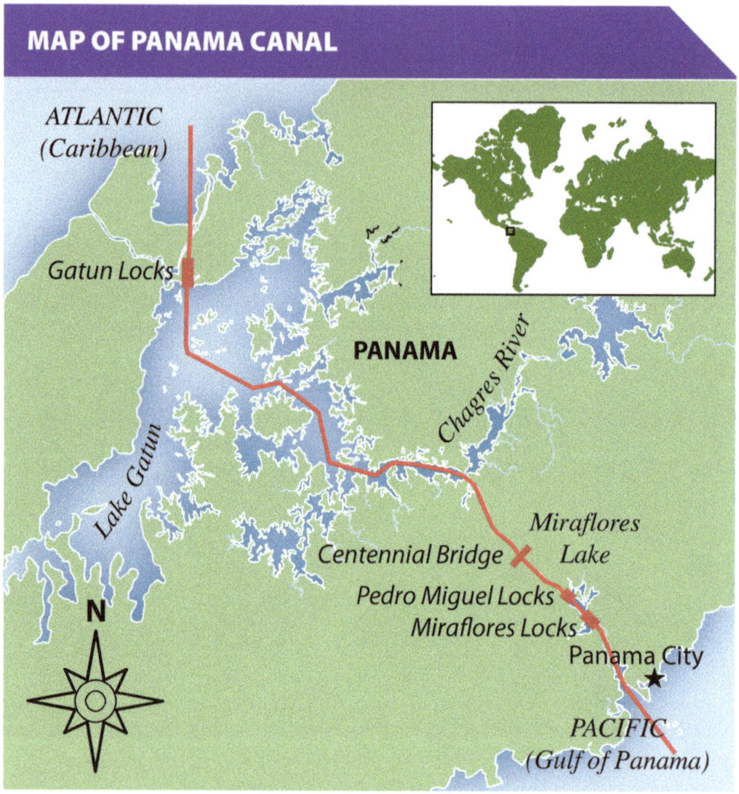

beauty, but the picturesque portion of the journey is reached when the vessel enters Culebra Cut. Here for the first time the spectator gains a conception of the tremendous task which has been accomplished.

It is not a full conception for he is sailing on a water level 45 or 47 feet above the bottom of the excavation which was made and the height of the banks is diminished to that extent.

Still as he passes between the towering walls of the mountain range, if he be a true American, he will feel a thrill of pride and patriotic enthusiasm at the work which his countrymen have accomplished.

Leaving the Culebra Cut the vessel advances to the first pair of locks on the Pacific side, at Pedro Miguel.

The vessel is tied up at the approach wall and is towed into the lock, as at Gatun, the water level in it being that of the lake.

When inside the lock the gates behind the ship are closed, and through the openings in the floor the water is let out until the level in the lock is that of the small lake below.

The vessel then passes into Lake Miraflores, a small body of water about a mile in extent, and proceeds under its own power till the first of the two locks at Miraflores is reached.

Here it enters the first lock, under the same conditions as prevailed at Pedro Miguel. The gates are closed and the water in the lock is let out into the second lock below until the two are brought to the same level. Then the vessel is towed forward into the second and the gates are closed behind it.

Finally, the water in the second is let out into the approach channel and the level in the hike is reduced to that of the sea. The gates are then opened in front of it, and it is ready to proceed into the Pacific ocean.

The time consumed in the journey ranges from 10 to 12 hours according to the speed of the ship.

The time consumed in passing the locks is three hours, one and a half hours in the three locks at Gatun and about the same time in the three on the Pacific side, or a half hour in each lock. . . .

The locks are brilliantly lighted by electricity in lamps grouped upon concrete poles 30 feet in height. The lamps are clustered under concrete hoods in such a manner as not to strike the eye while illuminating clearly all parts of the structures.

Along the line of the channel there are range lights of electricity except in the floating buoys marking the channel through Gatun Lake where acetylene gas is used.

The most powerful range lights, in high concrete lighthouses, are placed at the Atlantic and Pacific entrances. They will be visible for from 12 to 18 nautical miles at sea.

Worldwide Benefits

By opening a new ocean highway which will shorten great trade routes by thousands of miles, all nations will benefit, for the inevitable effect will be to cheapen transportation and reduce the price of commodities.

The new canal gives the continent of North America a continuous coast line from Labrador to Alaska, thus securing for the United States control of the commerce of the Western Hemisphere.

It reduces the distance between New York and San Francisco from 13,135 miles to 5,263—a saving of 7,873 miles.

It brings the west coast of South America, most of the large commerce of which is controlled by Europe, 3,000 miles nearer to the United States than to Europe, with the probable result of ultimately shifting control to the United States.

It will shorten the distance between New York and Yokohama [Japan] by more than 3,000 miles, between New York and Melbourne [Australia] by nearly the same distance, and between New York and Hong Kong by 245 miles. This will give the United States the advantage in competing with Europe for the commerce of Japan, and place it on equal terms with Europe so far as distance is concerned, in competing for the commerce of Australasia, China and the Philippines. . . .

> By bringing the nations of the earth into closer relations . . . the peace of the world will be more firmly secured.

Whatever else may be the effect of this new ocean highway, cheaper commodities for the world can safely be predicted. Direct competition along many great trade lines cannot fail to secure this result.

By bringing the nations of the earth into closer relations, and by developing the commerce of one with another and enhancing its value, the peace of the world will be more firmly secured, for the cost of wars, in disturbing and destroying commerce, will be an increasingly powerful influence against them. . . .

The Panama Canal is the supreme engineering feat of the ages.

It has been pronounced the greatest of the world's wonders of work.

It has been completed far ahead of the time fixed, for several million dollars less than the estimated cost, in a manner so excellent as to defy criticism, and without a shadow of scandal of any sort.

It is the work of Americans and the manner of its accomplishment has brought honor to the nation throughout the world.

It was a great administrative as well as a great engineering task, for the expenditure of $375,000,000 had to be regulated and guarded and a colony of 60,000 souls—the number of the canal workers and their dependents—had to be governed.

Success, complete and honorable, has been secured because from beginning to end the task has been in able and honest hands. Fitness was the sole test of service, and the baleful touch of sordid politics was not allowed to fall upon it.

VIEWPOINT 6

Expectations of the Influence of the Canal

Holland (Elisha Jay Edwards)

The following 1913 newspaper article reports what several eminent men have said about the expected impact of the Panama Canal. A French nobleman believes that it will help to establish world peace and feels that British and European commentators are more aware of its potential long-term benefits than Americans are. A British economist's address to the British Association for the Advancement of Science expresses a similar view. He correlates statements of leading American entrepreneurs about the canal's significance to an extent that no American speaker or writer has done, according to the author. Elisha Jay Edwards was a well-known journalist and newspaper editor who frequently wrote opinion pieces under the pen name "Holland."

SOURCE. Elisha Jay Edwards, "'Holland' Writes of Magnitude of Panama Canal's Influence," *Washington Post* archive, September 23, 1913. Courtesy of The Washington Post Writers Group.

The Panama Canal which your country is building will, I think, become a powerful influence in the permanent establishment of world peace." This statement was made to the writer by the Baron [Paul Henri] d'Estournelles de Constant [recipient of the 1909 Nobel Peace Prize] a few hours before his departure from the United States at the conclusion of a tour whose itinerary enabled him to deliver many addresses in various parts of the Union. The baron was too courteous to say that he feared the people of the United States did not begin to realize what the political, economic, commercial, and international influences which would be created by the opening of the Panama Canal to navigation were likely to be, but he clearly intimated that he thought the American people at present were concerned chiefly with the immediate commercial advantages to be established between the United States, the West Indies, and the west coast of South America. In fact, it seemed to be the impression of this distinguished French nobleman that a better insight into the potential influences of the Panama Canal had been obtained by leading men of Great Britain, France, and Germany, than by all save a very few in the United States.

British View of the Canal's Importance

This view will be regarded as having been confirmed to some extent by the address which was delivered a few days ago by Prof. Adam W. Kirkaldy before the British Association for the Advancement of Science. Prof. Kirkaldy's name is not a familiar one in the popular sense, but is known to men of science, and especially to the masters of economic science wherever the English language is spoken, and also in France and Germany. As the chief executive of the department of finance of Birmingham University he has attained high authority among scholars, particularly upon economic subjects.

After the War, the Canal Will Fulfill Its Promise

The Panama Canal was heralded in advance as the greatest economic blessing that the century could bestow upon the human race, as well as the most wonderful engineering feat of any age. A measure of oriental hyperbole greeted the prospective embrace of the waters of the oceans. New links of amity were to be forged. The East and the West were to be brought into glorious conjunction. A degree of common interest was to be established, through which misunderstandings arising from distance and strangeness would vanish in the presence of a growing intimacy that had in it nothing of contempt, but rather of increased respect.

All of which was more than justified. That a continent-wide war should supervene almost before the last spadeful of earth was thrown out by American hands had not then been reckoned with. That the development of the European struggle should have reached the point

The brief synopsis of his address, which was cabled to the United States, contains statements which in one way or another and by different American citizens have been made from time to time, especially recently. But no citizen of the United States has, at least in public address or publication, brought together so many important statements and reflections as are contained in the address to the British men of science by Prof. Kirkaldy. The late E.H. Harriman had a fairly good understanding of what some of the features of the commerce that will be floated through the Panama Canal are to be, yet Mr. Harriman's view was not so comprehensive as was that of the late C.P. Huntington, although Mr. Huntington died before work was actually begun upon the Panama Canal. Mr. Huntington was convinced that an artificial waterway across the Isthmus would tend greatly to the growth of the West Indian islands and that of the west coast of South America and of the gulf States of the Union. He was persuaded that commerce through the where a titanic battle between the contending forces was so imminent as to overshadow the opening of the canal could not have been foreseen.

Yet these things have come to pass. On the day that the canal was declared formally ready to serve as the common carrier of the peaceful argosies of the nations, the attention of the world is drawn to scenes of carnage. . . . Much that has been built up by slow and laborious effort is now being torn down in a moment. The victories of peace are obscured, the arts of war are regnant.

But the canal will endure. When the strife is over, it will be the artery through which will flow the revived commercial blood of the nations. When peace comes, the Panama Canal will enter into its own as a factor for good in the destinies of all the peoples of the earth.

SOURCE. Washington Post *archives, August 16, 1914.*

canal would not seriously impair railroad transportation, but would, in fact, tend greatly to increase it.

Benefits of the Canal to the West Indies, Central, and South America

Mr. Frank Trumbull, chairman of the Chesapeake and Ohio Railroad Company's board of directors, recently visited the Panama isthmus and studied the canal from the scientific and engineering point of view rather than from the commercial. Mr. Trumbull's investigation inclines him to think that the commercial business done through the canal may at first prove not to be so great as many persons have thought likely. He did intimate in a brief article written after his return that he could see stupendous and economic and political changes in American life ultimately beneficial to American prosperity, tending also to establish the position of the United States as the world's great power through the influences created by the Panama Canal.

> The nations of Central and South America . . . will find themselves growing rapidly and greatly.

What Mr. Trumbull vaguely hinted at in his short article is treated at considerable length by Prof. Kirkaldy. He foresees very great influences of the beneficial character upon the West Indian islands, venturing even to say that they will enter upon a period of new and vast prosperity through the kindly influences created by the canal. Furthermore, he asserts that the nations of Central and South America which are now almost asleep, not yet realizing what their stupendous resources and opportunities are, will find themselves growing rapidly and greatly, partly on account of immigration from Europe and partly on account of great increase of trade.

Inspired Flagler's Railway

That was the view taken by the late Henry M. Flagler and upon which he justified his daring and costly experiment of constructing a railroad upon the viaduct plan around the southern coast of Florida to a terminal at Key West. Men of capital who knew Mr. Flagler well and who admired him sometimes ventured to ask him if he had not uselessly risked great capital in the construction of his railroad and in the attempt to establish a terminal at Key West. He would reply in a few words, but these made it plain that he possessed the true powers of imagination which are characteristic of the ablest constructive business intellects. He pointed out that Key West was the true gateway of the Panama Canal, and that there would therefore radiate from the terminal there material influence which would vastly invigorate Cuba and create such prosperity upon that island as to be certain permanently to establish peace. These influences radiate from Key West to South America, in Mr. Flagler's opinion. This undoubtedly was the point of view taken by Baron d'Estournelles de Constant when he said that the influ-

The USS *Missouri* was the first warship to travel through the Panama Canal. (© Gamma-Keystone via Getty Images.)

ence of the Panama Canal would be great and might be preeminent in establishing permanent world peace.

Old Methods Must Go

Prof. Kirkaldy is quoted as having said that the chief effect of the opening of the Panama Canal to navigation, so far as the outside world is concerned, will probably be that it will cause commerce and trade to pass out of its present grooves, and for that reason will tend thoroughly to modernize business methods. There was a hint of this idea, it is thought, in some things which Mr. Flagler is quoted as having said, and, in all probability, the idea occurred to Mr. Trumbull while he was on the Panama isthmus last winter. There may have been a thought of this kind in the mind of C. P. Huntington when, in the last year of his life, he spoke of the powerful international influences which would be created by a canal across the isthmus. Prof. Kirkaldy goes into details to a greater extent than any American has done, for he is reported to have said that old methods of commerce will have to go to the scrap heap. Apparently he is familiar with a term which the manufacturers of the United States for some years have better understood than the manufacturers of Great Britain, who to this day have hardly overcome their feeling that the American scrap heap represents extravagance in manufacturing. Prof. Kirkaldy is reported to have said that friction among the factors of production would have to be eliminated. Those who have been instrumental in creating modern combinations or so-called trusts have based their philosophy on the beliefs that combinations of that kind will eliminate friction in the factors of production. The professor is inclined to think that one of the remoter consequences will be that capital and labor in competing countries will find it imperative that they work harmoniously together. This, he asserts, will bring about results of magnitude, both social and economic, and he asks if this is what America dreamed of when she entered on her stupendous undertaking. It is thought here that this eminent British authority's address will serve greatly to influence the American people to study their stupendous undertaking from all points of view.

[Due to the canal] capital and labor in competing countries will find it imperative that they work harmoniously together.

CHAPTER 2

Controversies Surrounding the Building of the Panama Canal

VIEWPOINT 1

The Role Played by the United States in the Panamanian Revolution Was Ethical

Philippe Bunau-Varilla

Photo on previous page: Laborers from Spain work on building the Panama Canal in 1909. Workers were brought from many countries to build the canal. (© **Bettmann/Corbis/AP Images.**)

Philippe Bunau-Varilla was the chief engineer of the French effort to build a canal in Panama. He was instrumental in precipitating the revolution that gained Panama's independence from Colombia and in persuading US president Theodore Roosevelt to send support ships. As a result, when Panama was recognized as a nation, he became its Minister to the United States. However, many criticized his actions because he had profited financially from Panama's sale of the canal rights, and many also claimed that the United States had instigated the revolution in order to get those rights. In the following excerpt from Bunau-Varilla's autobiographical book about Panama, he replies to those accusations.

SOURCE. Philippe Bunau-Varilla, *Panama: The Creation, Destruction, and Resurrection*. Ann Arbor: University of Michigan Library, 1914, pp. 302–303, 334–339. Courtesy of University of Michigan Library.

A grave question of conscience confronted me. "Had I the moral right to take part in a revolution and to encourage its development?" My answer was: "Yes."

"Yes, because I had twice warned President [of Colombia Jose Manuel] Marroquin in November 1902 and in June 1903, of the grave risks to which its anti-canal policy exposed Colombia.

"Yes, because I had again notified these risks to the Vice-President of the [Colombian] Senate, General Nel Ospina, in August 1903.

"Yes, because I had, for a fourth time, given the widest publicity to the fact in the article of the *Matin* on September 2, 1903.

"Yes, because Colombia was obviously prosecuting a policy of piracy aiming at the destruction of the precious work of Frenchmen.

"Yes, because this great enterprise was placed in one of those ill-defined moral zones where politics tyrannise over law and violate justice with impunity.

"Yes, because in the absence of law, and in presence of arbitrary political action, every man has the right to oppose another political action in order to bring about the triumph of justice."

I certainly had the moral right to annul, by political action at Panama, the fatal effect which political action at Bogota [the capital of Colombia] was bound to have on the gigantic French interests of which I was the sole defender.

The only objection that could have been made, might have been that I had taken the Colombians unawares. But I had four times individually or publicly shown to Bogota the consequences of its policy!

Colombia had refused to pay the slightest heed. She had declared war against law and justice. Nothing remained but to carry on the war. It was for me a cruel duty, but it was my duty.

My conscience could not reproach me if I backed the revolution and took the leadership of those who were disposed to risk their lives in defence of their country's interests.

I exposed myself to the violent attacks of the Colombians, but I assured the existence of the work which, later on, was to flood their country with prosperity. I should have to remain exposed for many years to mad accusations, but the day would come when Colombia herself would render me justice and recognise that I had been actuated by concern for her superior interest. Meanwhile I was sure to be approved by the civilised world. . . .

Help Arrives from the United States

The 2nd of November [1903] passed without any news. On the third day expired the last day of the period of one week which I had fixed for the revolution, after the arrival of [Manuel] Amador at Panama.

Deeply disturbed by this silence I went on the morning of the 3rd to the offices of M. Lindo. I wanted to prepare with his ordinary code a despatch [to Panama] which my conventional code did not allow me to send on account of its incompleteness. I wished to make a supreme appeal to the energy and courage of the people of the Isthmus. . . .

As I left the building to go to the telegraph office, a newsboy ran up to me and offered me the *Evening Telegram.* I bought it, and cast a glance at it.

It announced the landing of General [Juan] Tovar, and of the Colombian troops the very same morning at Colon [a Panama port city], as well as the arrival on the previous evening of the [US ship] *Nashville.* Nothing more—not a word of the slightest revolutionary movement.

Everything seemed to be irretrievably lost. . . .

I spent the whole afternoon in a state of profound dejection. My dear wife tried to comfort and encourage

Philippe Bunau-Varilla was instrumental in Panama being recognized as a nation and was also criticized for profiting from it. (© **Bettmann/Corbis/ AP Images.**)

me in this infinite sorrow. Finally, she prevailed on me to dominate my grief and to go with her to dine. . . .

When I returned to the Waldorf Astoria at about 10 o'clock that evening a telegram was handed to me. It was in plain language, and signed by Amador. It ran thus:

> "*Proclamada Independencia del Istmo sin sangre.* [Independence of the Isthmus proclaimed without bloodshed]—Amador."

The life of the great undertaking had been saved at the very moment when it seemed to be destroyed for ever.

What had taken place?

The rumour of the arrival of the American man of war that I had announced had promptly leaked out and spread all over the Isthmus.

From the morning of the 2nd November, all the inhabitants of Colon were looking towards Kingston [Jamaica], hoping for the appearance of the ship symbolising American protection.

As the hours passed disappointment gradually invaded all hearts.

Towards nightfall despair was general, when suddenly a light smoke arose in the direction of the north-east.

This was a ray of hope! If it were the liberator?

Little by little the smoke thickened, the ship emerged above the horizon and soon the Star-Spangled Banner dominated the Bay of Colon.

A burst of delirious enthusiasm shook the whole Isthmus.

> "The revolution had been made without shedding a drop of blood."

It was really true! Bunau-Varilla had effectively obtained for the unfortunate country the protection of the powerful Republic!

At this moment without one word having been uttered the Revolution was accomplished in the hearts of all. The *régime* of Colombian tyranny was over!

A Bloodless Revolution

The people were so intoxicated with joy, that serious business was postponed until the following day.

Instead of supplying the wharves of Colon with an armed force to prevent a possible landing of the Colombian troops, nothing was done.

This would have entailed the immediate interference of the American cruiser, and prevented a landing which would have provoked disorder. But the confederates had forgotten this detail in their blind happiness.

It happened that this arrival of the Colombian troops, which they had invented in order to justify the despatch, *Press steamer Colon,* took place the very same day they had announced.

In the morning of the 3rd November, General Tovar arrived quietly with about 500 soldiers.

It was the news which the *Evening Telegram* had brought to me.

This unexpected event awoke the confederates. The employees of the Panama Railroad availed themselves of various technical pretexts to delay the formation of a special train required for the troops.

General Tovar took the train for Panama, leaving his troops behind him at Colon.

Meanwhile the patriotic excitement determined by the arrival of the *Nashville* was steadily gaining on the entire population as well as the garrison of Panama.

> The Republic of Panama was born owing to American protection . . . [not] from a conspiracy fomented by the American authorities.

The aged Dr. Amador [who became the first president of Panama] gave the example. He went to the barracks, and started the whole movement by having General Tovar and his officers arrested by General [Esteban] Huertas, Commander of the Panama garrison.

The Independent Republic of Panama was immediately proclaimed.

The revolution had been made without shedding a drop of blood. It was due to the unanimous explosion of a whole nation, which refused to be stifled by the policy of Bogota.

But as it happens with nations depressed by long military oppression this explosion had taken place only when the people felt they were no longer alone.

This revolution, which it would have been so easy to accomplish from the 27th of October, when there were no obstacles in the way, was accomplished in face of the dreaded troops of the tyrant.

If these troops had arrived twenty-four hours earlier nobody would have made a move.

But they had landed twelve hours after the symbolic arrival of the *Nashville* had fired in all hearts the spark of hope, and thus restored general self-confidence. People had seen therein the extended hand of the powerful neighbour Republic. And this proof of friendship had made all hearts beat more quickly and raised everybody's courage.

The Republic of Panama had therefore been born; and it had sprung from a legitimate revolt against the most intolerable oppression.

The Revolution Was Not Fomented by the United States

Colombia can say to-day that the Republic of Panama was born owing to American protection. This is true if the word protection is understood as expressing solidarity between the mighty and the weak in the defence of common and legitimate interests. It was not born from a conspiracy fomented by the American authorities. It developed out of the simultaneous and parallel, but distinct, movement in two separate spheres of the same aspiration, the completion of the Panama Canal. Every one remained in his proper place and acted his legitimate part.

> The action of President Roosevelt was as correct as it was active and resolute.

Mr. [US president Theodore] Roosevelt avoided, during the first revolutionary attempts, anything which could resemble collusion. The abandonment of Amador by those who had promised him everything is the obvious demonstration that the American Government had refused to lend itself to anything of a compromising character.

The action of President Roosevelt was as correct as it was active and resolute. . . .

The claim of Colombia is, and will remain, untenable, because she herself forfeited her right by her policy.

Her rights challenged superior rights: the right of a nation to exist; the right of humanity to circulate.

She violated the very basis of her sovereign rights, namely, the duty of the sovereign to protect his subjects.

With a stroke of the pen [that is, the refusal to ratify a treaty with the United States for purchase of canal rights] she condemned the whole of the population of one of her provinces to destruction.

With a stroke of the pen she challenged the whole of humanity which had a pre-eminent right of way across the Isthmus.

With a stroke of the pen she cynically announced her will to confiscate from the French share and bond holders all that still remained from the wreck of the great enterprise.

With a stroke of the pen she disavowed her contract for the extension of the term of the French concession, on the pretext that certain formalities had not been fulfilled, whereas through her own fault, it was a physical impossibility to fulfil them.

These are the violations of superior rights which made the revolution of Panama the most legitimate of protests against tyranny. . . .

The Reaction in America

On the day following, November 2, was published [in the *New York Sun*] the admirable article of which I give the following extracts.

> COLOMBIA BEFORE THE WORLD TO-DAY
>
> The Congress at Bogota adjourned on Saturday without having ratified the Canal Treaty negotiated by the Colombian Government with ours. The end of the session

opens no hopeful path toward an agreement hereafter, with these same politicians or their successors. As all the world knows, the terms of compensation offered by the United States in the Treaty were liberal beyond precedent; the contemptuous rejection of these terms by the [US] Congress, after their acceptance as satisfactory by the representatives of Colombia in the preliminary process, has had no explanation that is respectable, and but one explanation that is credible. That is blackmail; attempted blackmail of the French owners of the canal property, because it was believed at Bogota that they would submit rather than lose the opportunity to sell for forty millions; attempted blackmail of the United States, because it was believed at Bogota that this rich and enterprising nation would pay to the very limit of tropical avarice rather than give up its project or take the alternative of the vastly inferior route through Nicaragua. . . .

The attitude of Colombia as an obstructionist for extortion is the same toward the whole civilised world as it is toward the United States in particular. That Government is trying to block the mightiest enterprise ever undertaken for the common benefit of the globe's commerce. Colombia's duty to the world is repudiated along with her obligations to the United States. She is neither entitled to the sympathy and moral support of any great power, nor likely to receive such sympathy and support.

VIEWPOINT 2

The Role Played by the United States in the Panamanian Revolution Was Not Ethical

Abelardo Aldana

The Colombian writer of the following viewpoint published in 1904 expresses his anger at the actions of the United States in instigating Panama's revolt against Colombia. He says that the revolution would not have happened without US aid—that it was not a real revolution wanted by the Panamanian people, but simply a conspiracy by a few men who profited from the sale of the canal rights. He argues that the United States' support was a violation of the 1846 treaty with Colombia that secured neutrality with respect to the Isthmus of Panama and Colombia's sovereignty over it. The United States has behaved dishonorably and is guilty of international lawlessness, he says; if it does not redress the wrong done to Colombia, it will be considered untrustworthy throughout the

---

SOURCE. Abelardo Aldana, *The Panama Canal Question: A Plea for Colombia*. Wales: Cardiff, Western Mail, Limited, 1904, pp. 12–18.

civilized world. Abelardo Aldana was the consul of Colombia at Cardiff, Wales.

Let us consider now the action of the United States Government in connection with the conspiracy at Panama.

The rejection of the Herran-Hay treaty by the Colombian Government caused a conspiracy which broke out in the city of Panama on the 3rd of November last [1903], which, being aided by the National guard stationed there (by bribery, as no other motive could have induced the officers and soldiers to betray their country), they imprisoned the authorities, established a so-called government of three consuls, and sent an emissary to Washington. The United States Government, who were undoubtedly ready for action, immediately despatched men-of-war to Colon [a Panama port city] and Panama, with instructions to prevent the landing of the Colombian troops which might be sent to re-establish order. Within 48 hours this new government was formally recognized by the United States, and just three weeks after the revolt a treaty was signed. . . .

This simple narrative of the facts, taken in conjunction with the terms of the treaty, clearly shows the complicity of the United States Government in the revolt, if not in the actual promotion of it, which could not have happened without the knowledge that it would receive all the necessary assistance for its maintenance, and which would not have lasted beyond 48 hours without the help rendered by the American forces at Colon. . . .

> No political programme or social reforms have been proclaimed by the conspirators.

We must bear in mind that no political programme or social reforms have been proclaimed by the conspirators, and that which more clearly shows the nature and the true aim of the revolt is that no mention has been made of a general election in the affected territory for the establishment of a regular and formal Government. It is simply monstrous that a Treaty, which practically annexes the territory of the Isthmus to a foreign Government, may not have been considered by the duly elected representatives of the inhabitants whose birthright is thus lightly disposed of. But of what value are rights and formalities to a Government sinfully conceived and treacherously brought forth for a mercenary purpose, the spoliation and dismemberment of the Colombian territory in exchange for £2,000,000. Apart from the connivance of the American Government with the spoliators of the territory of a friendly Nation, the recognition of the revolutionary Government of Panama, and the assistance given to it, is an evident violation of the international equity prevailing between civilized nations, because it is a paramount duty to be neutral in the domestic affairs of another State.

The United States Violated Its 1846 Treaty with Colombia

Besides, this action of the United States Government is a flagrant breach of good faith, inasmuch as it annuls the guarantee given by the Treaty, signed at Bogota on the 12th December, 1846, between the Government of New Granada [Colombia] and the American Minister. The Article XXXV of this Treaty says:—

> The United States of America and the Republic of New Granada desiring to make as durable as possible the relations which are to be established between the two parties by virtue of this Treaty, have declared solemnly and do agree to the following points:—
>
> The United States guarantee *positively and efficaciously* [italics are the author's] to New Granada by the present stipulation the perfect neutrality of the before-mentioned Isthmus, with the view that the free transit

> from the one to the other sea may not be interrupted or embarrassed in any future time while this treaty exists; and *in consequence*, the United States also guarantee in the same manner the rights of *sovereignty* and *property* which New Granada has and possesses over the said territory.

This solemn pledge was emphatically confirmed by Mr. [US Secretary of State William] Seward, who said that it meant that "if the complete and absolute sovereignty and independence should ever be assailed by any Power *at home* or abroad, the United States will be ready, co-operating with the Government and their ally, to defend them." Several other declarations made by American Presidents, Secretaries of State, and international representatives, stand to prove that the policy of the United States has always tended to reassert its pledge of maintaining the sovereignty and property of Colombia on the Isthmian territory. . . .

The United States Government is by this public document the guarantor to Colombia of her property and sovereignty on the territory of the Isthmus, and is in honor bound to keep its solemn promise loyally; but instead of honorably discharging its duty, it has recognized the plunderers of the territory she undertook to guarantee, and is treating with them as the rightful owners, and is acting thus, expecting to obtain at least a substantial portion of the spoil. This conduct is so discreditable that it far surpasses the most atrocious spoliations which have disgraced the history of even unscrupulous conquerors, because a guarantor who becomes an accomplice expecting to share the plunder is something worse than the most unprincipled conqueror. . . .

Mr. [US Secretary of State John] Hay speaks of the duty of maintaining undisturbed the traffic across the Isthmus as the excuse for the intervention in the domestic affairs of Colombia. This traffic had not been dis-

US Secretary of State John Hay brokered the Herran-Hay treaty, which was rejected by the Colombian government. (© **Hulton Archive/ Getty Images.**)

turbed when the revolution broke out, and the only cause which might disturb it is the breach of the legal order which has taken place. So the straightforward manner to discharge the alleged duty should have been to assist, if required by the lawful authority, in suppressing the disturbing element. . . .

The Revolution Was Not a Political Movement

[US] President [Theodore] Roosevelt, addressing the emissary of the so-called Republic of Panama, said he had "seen in the events that had recently occurred an

> The revolution was brought forth . . . merely as a deal to appropriate Colombian rights.

unopposed expression of the will of the people of Panama." If he referred to the people of that city he may be right; but he is wrong if he meant the will of the people of the Department, which has had no part in the conspiracy hatched by some foreign adventurers in the city of Panama, assisted by some interested people at New York and Washington, and, perhaps, a few unworthy Colombians. The revolution was brought forth, not as a political movement for breaking a tyrant's yoke, but merely as a deal to appropriate Colombian rights to sell them to a foreign Power and to divide the plunder. The President said also that the ancient territory of Panama is reasserting its right of self-control, implying that the said territory had had this right before, which tacit assertion is inexact. He appears very zealous for the right of popular government, and stretches out the hand of fellowship towards the infant State; but he ignores the rights of a sister Republic, which might have received a little more consideration if she had been in a position to defend herself against the unwonted aggression from the so-called protector. What would the President say if the standard of independence were raised in Florida or Maine, and another Power rushed in before a week was out with recognition of the "new Republic"? . . . In his previous message to Congress President Roosevelt says that if Colombia did not come to an agreement "we must forthwith take the matter into our own hands." This frank determination to disregard the rights of other Nations when they are in the way of American interests, is a danger to peace and civilization, and a measure of the trustworthiness of a Government which breaks its promises to suit its convenience. . . .

A "moral trusteeship" is the descriptive term for the treaty between America and the new Republic, which implies that the Canal will be constructed for the benefit of the commercial world, and the United States will act simply as the guardian of it. The Canal is merely an American political and commercial concern, and the proclaimed neutrality will be subservient to the interests and conveniences of the United States. Other Nations may benefit by it, provided their interests will not clash with those of America, but never otherwise, and to this extent the meaning of this "moral trusteeship" must be limited. . . .

The United States Is Guilty of International Lawlessness

I think what I have stated is sufficient to convince any unbiassed mind that the American Government has abetted, if not directly promoted, the conspiracy of Panama, and that it has flagrantly violated the treaty of 1846. It stands, therefore, before the conscience of the civilized world, guilty of international lawlessness and perfidy.

> The American government . . . [is] guilty of international lawlessness and perfidy.

The consummation of the attempt which is being made by the conspiracy at Panama would produce these results: The Colombian territory as it has existed since the formation of the nation would be dismembered; Colombia would irreparably lose, besides her dominion rights on the Isthmus, £2,000,000 in cash and an annuity of £50,000, which was the price offered by the United States Government for the concession, and this would be "monstrously unfair," as *The Financial News* qualifies the "filching from her not merely territory, but territorial rights of great value."

On the other hand, the honor of the United States would be tarnished and her reputation blasted. Henceforth it would not appear as the protector, but as the perfidious enemy of the Latin-American States, and its commerce would suffer considerably. Moreover, the

American Government would be morally, if not legally, bound to pay the Colombian debts, inasmuch as it would deprive her of her most valuable asset.

Colombia would readily accept a Treaty with the American Government for the construction of the canal on honorable and satisfactory conditions for both countries. Then America could accomplish the great enterprise with a pure conscience and clean hands. The fatherland of Washington would not descend from the lofty plane on which it stood when it faced the Old World as the sincere protector of the Hispano-American States.

We have appealed to the Latin-American States, to the people and Senate of the United States, to England, and to the whole of the civilized world; but as yet Colombia seems powerless and friendless. Our best hope of obtaining redress for the cruel wrong done to her is in the Senate of the United States, who, we trust, will be swayed by the command: "Let judgment run down as waters, and righteousness as a mighty stream."

VIEWPOINT 3

The Actions of the US President in Acquiring the Canal Zone Were Justifiable

Theodore Roosevelt

In the following excerpt from his autobiography, Theodore Roosevelt explains his support of the Panamanian revolution. He declares that he exercised his authority in sending armed ships to Panama because he could see that Congress would debate indefinitely about acquiring the canal rights, and the time for action had come. The Panamanians had attempted to revolt many times in the past, and the United States had defended the Isthmus at Colombia's request. Now Colombia was ruled by a dictator and had refused to ratify a treaty giving the United States the canal rights. No one in the US government had incited or encouraged the revolution, he says; he intervened merely to prevent Colombia from suppressing Panama's revolution. In his opinion this was a moral action

---

SOURCE. Theodore Roosevelt, *Theodore Roosevelt: An Autobiography*. New York: Palgrave Macmillan, 1914, pp. 553–571.

in the best interests of the United States, Panama, and the world at large. Roosevelt was the twenty-sixth president of the United States, serving from 1901 to 1909.

By far the most important action I took in foreign affairs during the time I was President related to the Panama Canal. Here again there was much accusation about my having acted in an "unconstitutional" manner—a position which can be upheld only if Jefferson's action in acquiring Louisiana be also treated as unconstitutional; and at different stages of the affair believers in a do-nothing policy denounced me as having "usurped authority"—which meant, that when nobody else could or would exercise efficient authority, I exercised it.

During the nearly four hundred years that had elapsed since [Spanish explorer Vasco Núñez de] Balboa crossed the Isthmus, there had been a good deal of talk about building an Isthmus canal, and there had been various discussions of the subject and negotiations about it in Washington for the previous half century. So far it had all resulted merely in conversation; and the time had come when unless somebody was prepared to act with decision we would have to resign ourselves to at least half a century of further conversation. Under the Hay-Pauncefote Treaty signed shortly after I became President, and thanks to our negotiations with the French Panama Company, the United States at last acquired a possession, so far as Europe was concerned, which warranted her in immediately undertaking the task. It remained to decide where the canal should be, whether along the line already pioneered by the French company in Panama, or in Nicaragua. Panama belonged to the Republic of Colombia. Nicaragua bid eagerly for the privilege of having the States build the canal through her territory. As long as it was doubtful which route we would decide upon, Colombia extended every promise of friendly coöperation: at the [1901–1902] Pan-American Congress in Mexico her delegate joined in the unanimous vote which requested the United States forthwith to build the canal; and at her eager request we negotiated the Hay-Herran Treaty with her, which gave us the right to build the canal across Panama. A board of experts sent to the Isthmus had reported that this route was better than the Nicaragua route, and that it would be well to build the canal over it provided we could purchase the rights of the French company for forty million dollars; but that otherwise they would advise taking the Nicaragua route. . . .

> [The United States] had again and again been forced to intervene to protect the transit across the Isthmus.

We had again and again been forced to intervene to protect the transit across the Isthmus, and the intervention was frequently at the request of Colombia herself. The effort to build a canal by private capital had been made under [French developer Ferdinand] De Lesseps and had resulted in lamentable failure. Every serious proposal to build the canal in such manner had been abandoned. The United States had repeatedly announced that we would not permit it to be built or controlled by any old-world government. Colombia was utterly impotent to build it herself. Under these circumstances it had become a matter of imperative obligation that we should build it ourselves without further delay. I took final action in 1903. During the preceding fifty-three years the Governments of New Granada and of its successor, Colombia, had been in a constant state of flux; and the State of Panama had sometimes been treated as almost independent, in a loose Federal league, and sometimes as the mere property of the Government at Bogota [the Colombian capital]; and there had been innumerable appeals to arms, sometimes for adequate, sometimes for inadequate, reasons. . . .

US president Theodore Roosevelt tests a steam shovel at the Culebra Cut section of the Panama Canal in 1906. (© **AP Images.**)

Revolution Had Often Been Attempted in Panama

Revolutions, rebellions, insurrections, riots, and other outbreaks that occurred during the period in question . . . number fifty-three for the fifty-three years, and they showed a tendency to increase, rather than decrease, in numbers and intensity. One of them lasted for nearly three years before it was quelled; another for nearly a year. In short, the experience of over half a century had shown Colombia to be utterly incapable of keeping order on the Isthmus. Only the active interference of the United States had enabled her to preserve so much as a semblance of sovereignty. Had it not been for the exercise by the United States of the police power in her interest, her connection with the Isthmus would have been sundered long before it was. . . .

Colombia was under a dictatorship. . . .

On July 31, 1900, the Vice-President, [José Manuel] Marroquin, executed a "coup d'état" by seizing the person of the President, [Manuel Antonio] Sanclamente, and imprisoning him at a place a few miles out of Bogota. Maroquin thereupon declared himself possessed of the executive power because of "the absence of the President"—a delightful touch of unconscious humor. He then issued a decree that public order was disturbed, and, upon that ground, assumed to himself legislative power under another provision of the constitution; that is, having himself disturbed the public order, he alleged the disturbance as a justification for seizing absolute power. Thenceforth Maroquin, without the aid of any legislative body, ruled as a dictator. . . . When the people of Panama declared their independence in November, 1903, no Congress had sat in Colombia since the year 1898, except the special Congress called by Maroquin to reject the canal treaty, and which did reject it by a unanimous vote, and adjourned without legislating on any other subject. The constitution of 1886 had taken away from Panama the power of self-government and vested it in Colombia. The *coup d'état* of Maroquin took away from Colombia herself the power of government and vested it in an irresponsible dictator.

Consideration of the above facts ought to be enough to show any human being that we were not dealing with normal conditions on the Isthmus and in Colombia. We were dealing with the government of an irresponsible alien dictator, and with a condition of affairs on

> People on the Isthmus . . . unanimously repudiated the Colombian government, and demanded that the United States build the canal.

the Isthmus itself which was marked by one uninterrupted series of outbreaks and revolutions. As for the "consent of the governed" theory, that absolutely justified our action; the people on the Isthmus were the "governed"; they were governed by Colombia, without their consent, and they unanimously repudiated the Colombian government, and demanded that the United States build the canal.

Colombia Had Agreed That the United States Should Build the Canal

I had done everything possible, personally and through Secretary [of State John] Hay, to persuade the Colombian Government to keep faith. Under the Hay-Pauncefote Treaty, it was explicitly provided that the United States should build the canal, should control, police and protect it, and keep it open to the vessels of all nations on equal terms. We had assumed the position of guarantor of the canal, including, of course, the building of the canal, and of its peaceful use by all the world. The enterprise was recognized everywhere as responding to an international need. . . . The United States had assumed in connection with the canal certain responsibilities not only to its own people but to the civilized world which imperatively demanded that there should be no further delay in beginning the work. The Hay-Herran Treaty, if it erred at all, erred in being overgenerous toward Colombia. The people of Panama were delighted with the treaty, and the President of Colombia, who embodied in his own person the entire government of Colombia, had authorized the treaty to be made. But after the treaty had been made the Colombia Government thought it had the matter in its own hands; and the further thought, equally wicked and foolish, came into the heads of the people in control at Bogota that they would seize the French Company at the end of another year and take for themselves the forty million dollars which the United States had agreed to pay the Panama Canal Company. . . .

By my direction Secretary Hay, personally and through the Minister at Bogota, repeatedly warned Colombia that grave consequences might follow her rejection of the treaty. . . . From the information I received, I deemed it likely that there would be a revolution in Panama as soon as the Colombian Congress adjourned without ratifying the treaty, for the entire population of Panama felt that the immediate building of the canal was of vital concern to their well-being. Correspondents of the different newspapers on the Isthmus had sent to their respective papers widely published forecasts indicating that there would be a revolution in such event.

Moreover, on October 16, at the request of Lieutenant-General [Samuel B.M.] Young, Captain Humphrey and Lieutenant Murphy, two army officers who had returned from the Isthmus, saw me and told me that there would unquestionably be a revolution on the Isthmus, that the people were unanimous in their criticism of the Bogota Government and their disgust over the failure of that Government to ratify the treaty. . . . (Accordingly I directed the Navy Department to station various ships within easy reach of the Isthmus, to be ready to act in the event of need arising.)

These ships were barely in time. On November 3 the revolution occurred. Practically everybody on the Isthmus, including all the Colombian troops that were already stationed there, joined in the revolution, and there was no bloodshed. . . . On the Pacific side a Colombian gunboat shelled the City of Panama, with the result of killing one Chinaman—the only life lost in the whole affair.

No one connected with the American Government had any part in preparing, inciting, or encouraging the revolution, and except for the reports of our military and naval officers, which I forwarded to Congress, no one connected with the Government had any previous knowledge concerning the proposed revolution, except such as was accessible to any person who read the newspapers and kept abreast of current questions and current affairs. By the unanimous action of its people, and without the firing of a shot, the state of Panama declared themselves an independent republic. . . .

Every consideration of international morality and expediency, of duty to the Panama people, and of satisfaction of our own national interests and honor, bade us take immediate action. I recognized Panama forthwith on behalf of the United States, and practically all the countries of the world immediately followed suit. The State Department immediately negotiated a canal treaty with the new Republic. . . .

From the beginning to the end our course was straightforward and in absolute accord with the highest of standards of international morality. Criticism of it can come only from misinformation, or else from a sentimentality which represents both mental weakness and a moral twist. To have acted otherwise than I did would have been on my part betrayal of the interests of the United States, indifference to the interests of Panama, and recreancy to the interests of the world at large. . . .

I did not lift my finger to incite the revolutionists. The right simile to use is totally different. I simply ceased to stamp out the different revolutionary fuses that were already burning.

VIEWPOINT 4

The Actions of the US President in Acquiring the Canal Zone Were Not Justifiable

James T. Du Bois

In the following 1914 article from the *New York Times*, an American diplomat expresses his dismay at the United States' support of the Panamanian revolution. In his opinion it was an injustice to Colombia, which he says was not ruled by a dictator as US President Theodore Roosevelt said. He points out that only a few of the Panamanians even knew about the revolution until the US Army proclaimed its success. He declares that many of Roosevelt's statements in his autobiography are wrong, although he does not question Roosevelt's motives. He argues in favor of the pending treaty that will compensate Colombia for the loss of Panama. (Roosevelt is referred to as Colonel, the military rank he held before becoming president and by which he was addressed after leaving office.) James T. Du Bois served as US minister to Colombia.

SOURCE. James T. Du Bois, "Du Bois Takes Issue with Roosevelt," *New York Times* archive, July 2, 1914. Courtesy of The New York Times Company.

I was in the Department of State in 1903 when the press announced that the mailed hand of the United States had stopped Colombian soldiers from suppressing a revolution on the Isthmus of Panama. Amazed at this enterprise, I said to those present: "This, if true, is a shocking injustice to Colombia, and the United States must pay dearly for it in the end."

Three years ago, when I was asked to go to [the Colombian capital] Bogota as Minister, I frankly stated my sincere sympathies for Colombia and was informed that friendly relations with that country were desired by the United States and that my sympathies were no bar to my mission.

An impartial investigation at Bogota, running over a period of two years, confirmed my sympathies and convinced me that instead of "blackmailers" and "bandits" the public men of Colombia compare well with the public men of other countries in intelligence and respectability, while the social life is as refined and cultured as can be found in any capital in the world—Bogota is called the Athens of South America.

Impressed by the humiliation and sufferings that had befallen Colombia by the act of 1903, I have worked unselfishly for the true interests of the United States in Latin America by trying to secure a settlement of our unfortunate differences on a broad and just basis, and the new treaty [compensating Colombia for the loss of Panama] will accomplish that result. I deeply regret the antagonism to it displayed in certain quarters and I deplore Col. [Theodore] Roosevelt's bitter and misleading attack embodied in his recent [1914] public statement.

> A handful of men, who were to be the direct beneficiaries of the revolution, conceived it, and not the hundredth part of the inhabitants of the Isthmus knew of the revolt.

He says: "The people of Panama were a unit in demanding the revolution." I say, and can prove it, that a handful of men, who were to be the direct beneficiaries of the revolution, conceived it, and not the hundredth part of the inhabitants of the Isthmus knew of the revolt until an American officer, in the uniform of the United States Army, raised the flag of the new republic.

An Act of War

"We never fired a shot at any Colombian," says Col. Roosevelt. True, but the execution of the orders of President Roosevelt to the American warship was an act of war, and if the Colombian soldiers had attempted to suppress the conspiracy they would have been captured, or driven into the sea or killed, because that is the way our brave marines have of doing things when they are let "loose."

Again, he says "his" attitude was "absolutely in accordance with the principles of the highest international morality." Col. Roosevelt's useful life challenges the admiration of men, but he is not infallible. If armed interference that prevents a friendly nation from maintaining sovereignty over her most valuable province, which we had pledged our national honor by solemn treaty obligations to maintain, is the "highest international morality," the lowest must be amazing, and 99 per cent. of the people who know the facts will support this view.

"France would have seized the Isthmus" and "the canal would never have been dug if I had not acted as I did," says Mr. Roosevelt. Nothing of record proves either statement; both are guesses, and both are simply multiplying suppositions by imagination. It is a matter of record that Colombia never seriously intended to seize the French company's property, and everybody knows that Colombia wanted the canal dug and wanted the United States to dig it, and had urged it for fifty years.

Everybody knows that Colombia wanted the canal dug and wanted the United States to dig it, and had urged it for fifty years.

Col. Roosevelt unfortunately feels that the new treaty discredits his administration in an act which he considers the most important of his life.

No self-respecting man will tolerate for a moment the thought that a single act of Col. Roosevelt's useful life was inspired by tainted motives, but no man is always right. In the Panama incident, while doing what he believed to be a great thing for mankind, and in accordance with the principles of the highest international morality, he did a bad thing for Colombia. He put himself in the position of the dentist apprentice who pulled the wrong tooth; he cannot put it back, and he does not want his employers to realize his mistake.

The President of Colombia Was Not a Dictator

Col. Roosevelt says that Acting President [Jose Manuel] Marroquin was an "absolute and uncontrolled dictator." He was a helpless and hapless old man placed in the midst of great national unrest and disorder at the close of the most disastrous civil war in the history of Colombia. He was not in accord with the [Colombian] Congress and the Congress was alive with discord, and 5,000,000 of innocent and hapless people saw the true interests of their country destroyed by the selfish disputes of less than one hundred men.

If Theodore Roosevelt had realized the true situation in Bogota at that hour he would have reinforced his patience with sympathy for that helpless people who had been, for a hundred years, our best friends south of the Rio Grande. But he was badly informed and cabled to Bogota, "The treaty must not be modified or amended," a right which all nations claim, and then placing the last straw on the back of Colombian pride, cabled again, "Ratify the treaty or you will regret it." The Colombian Congress, amazed and indignant, adjourned without considering the treaty.

Crowds gather to see the tugboat Gatún enter the Panama Canal in 1913. The Hay-Bunau-Varilla Treaty of 1903 gave the United States the right to build the canal and to control the area surrounding it. (© **Time & Life Pictures/Getty Images.**)

Within a short time the United States was urged by Colombia to reopen negotiations, but Col. Roosevelt concluded that "it was after the Fair" and immediately took Panama, commenced digging the canal, and let [the US] Congress debate the incident at leisure. He now opposes the new treaty on the grounds that it discredits his act, and he calls it a blackmailing agreement. . . .

The [US President Woodrow] Wilson Administration has taken a broader and more correct view of the entire question and has presented to the American people a treaty that ought to be heartily approved, because it is just to Colombia, and will remove forever an intolerable situation. Let us see what the treaty really means. . . .

Colombia Is Entitled to Compensation

As the third article is the most important feature of the treaty I consider it last. This article has been placed in the "Blackmailers' Club" by Col. Roosevelt with some harshness. He says that we might as well give Colombia forty millions as twenty-five millions, because she is not entitled to a cent. Is he right?

Under the contract of 1867 Colombia ceded the Trans-Isthmian Railroad Company to the Panama Railroad Company for a period of ninety-nine years, and for this she was to receive $250,000 annually during the life of the contract. . . .

The physical value of the railroad, shortly after the revolt, was placed at $16,446,000. In addition Colombia claims that she is entitled to several millions of dollars for that part of her national debt incurred in the interests of Panama before the revolution, a claim which was recognized in the treaty of 1909 made under the Roosevelt Administration. Altogether these claims total about $36,000,000. But ex-President Roosevelt insists that all of the civil rights, titles, and pecuniary interests of Colombia passed to Panama at the moment of actual separation, and therefor the third article of the treaty is a direct attempt to blackmail the United States.

The truth is the ownership and these rights had been virtually recognized by the Roosevelt Administration itself. . . .

In the spirit of justice, therefore, the United States, now in undisputed possession of the Panama Railroad, should compensate Colombia for the annuities which she is entitled to receive and also for the reversionary interests in the railroad. To say that these lawful interests passed to Panama at the separation is as unjust as it would be to say that we have discharged our obligations to one man by paying what he was lawfully entitled to into the hands of another man.

Many Americans Believed the Panamanian People Did Not Start the Revolution

In Panama the masses of the people not only did not know about the revolution until it had passed, but no more than an ordinary mob, such as may be aroused on an hour's notice in any city, participated in it.

It was not necessary that the people of Panama should know about it. The United States had agreed to stand between the clique of Panaman financiers and any offensive act Colombia might undertake. Undoubtedly there had been popular uprisings against Colombia in Panama, but the revolution of November 3, 1903 was not one of them. This revolution had three sources of inspiration: the French Canal Company, the capitalist Junta in Panama, and Theodore Roosevelt's desire to get a canal started before his inherited administration [ended]. . . .

The rightful owner of the territory we desired for a canal was Colombia. When we took that territory we took it from Colombia. The way we took it was to participate in a bogus revolution, engineered by a Junta of wealthy Panaman business and professional men. It turned out that the part they played in making the revolution a success was farcical, while the part the United States Marines played was vital. . . .

Anyone who will spend a month in Panama will discover that the republic would not stand from supper until breakfast if it were not for the supporting arm of the United States. It has become rather a burdensome task, too, as our interference three times with Marines to keep the government from toppling over proves. This is not because the Panamans are inferior to any other Central American peoples. It is because there is not sufficient inherent vitality in so tiny a republic to hold it up alone.

If any American railroad should desire property for a right of way and, instead of condemning it by due process of law, should connive with a neighbor to falsely claim possession of the property and then buy the property from the illegal owner, the action not only would not stand in law but it would outrage public opinion. That precisely is the course we followed at Panama.

SOURCE. *William R. Scott,* The Americans in Panama. *New York: Statler, 1912, pp. 241–243. www.openlibrary.org.*

These and other material claims of Colombia justify the indemnification article of the treaty, to say nothing of the tremendous territorial loss and humiliation that Colombia has sustained.

The time is not distant when Latin America will have a hundred millions of people inspired by new conditions of national and commercial life. Those now living feel that the Panama incident is the only real injustice committed by the United States against the Latin-American people. The treaty will correct that feeling and greatly change the sentiment that is now running heavily against us in all South America and place this country and Colombia upon that friendly footing so greatly desired by the people of both nations.

If any person believes it is pleasant for me to oppose a great leader whose fortunes I devotedly followed for ten years, and for whom I have profound esteem, he is radically mistaken. I am a Republican, and have been all my life, and I have been urged not to make this statement public because a successful issue of the treaty will help the Wilson Administration. I do not care to live to greet that day where my love of party smothers my love of justice and halts my courage in doing what I believe is right for the true interests of my country.

VIEWPOINT 5

The Panama Canal Construction Workers Are Treated Fairly

Robert M. Burnett and John V. Farwell Jr.

In 1907 a large group of US businessmen—members of the Commercial Clubs of major cities—chartered a ship and went to Panama for the purpose of examining the work being done on construction of the canal. According to the *Panama Star and Herald*, they were "in active management or control of some of the largest and most representative business organizations in the United States." They divided themselves into eight committees to investigate various aspects of the work, and aboard the ship during the return trip they held a general meeting at which the committee reports were presented. The following is excerpted from the reports of the committees on Housing and Food, chaired by Robert M. Burnett, and Efficiency of Labor, chaired by John V. Farwell Jr. Their conclusion was that working conditions were excellent, and they commended "the sympathy with the work and the zeal and intelligence of the men with whom [they] came directly in contact."

SOURCE. Robert M. Burnett and John V. Farwell Jr., *A Trip to Panama*. St. Louis: Lesan-Gould, 1907, pp. 142–149. Courtesy of Lesan-Gould.

For the group on Housing and Food, Mr. Burnett, as Chairman, presented the report, as follows:

The Subsistence Department on the Isthmus furnishes supplies of first-class quality at reasonable prices. . . .

There are no restrictions as regards diet. The purchaser of food from the Subsistence Department has the choice of the kind of food and the quantity he desires. In the case of the negro laborer, thirty cents per day is deducted from his pay for his meals, which are furnished him cooked, from the government kitchens. This is done to compel the negro to take sufficient nourishment as it was found he economized in his food at the expense of his health and strength.

The United States has built at Cristobal [a port in the Atlantic side of the canal] an up-to-date bakery with a capacity of from 20,000 to 30,000 loaves daily, and bread of first-class quality is furnished therefrom.

There are three grades of meals:

First. A thirty cents meal, served to Americans, where table-cloths, napkins and glassware are provided. This meal is served by waiters in the hotels. It consists of the following dishes:

Breakfast: Fresh fruit, cereals, bacon and eggs, potatoes, batter cakes, tea and coffee.

Dinner (at noon): Soup, an entree, roast beef, pork or mutton, (chicken twice a week), two vegetables and dessert.

Supper: Cold meats, canned or dried fruits, tea and coffee, bread and butter.

Second. The forty cents per day meal to Europeans, consisting principally of rancho stew made of meat, beans, potatoes and macaroni, with bread, tea and coffee.

Third. The thirty cents per day meal to negroes, consisting of bread, potatoes or yams, rice, beans, fresh or cured meat or salt fish, tea and coffee.

An ample supply of good drinking water is available at all times from the government reservoirs.

There are about 3,000 Americans and 3,000 Europeans who are fed at hotels and mess-halls, and 7,000 negro employees who are fed from the government kitchens, making a total of 13,000. There are 15,000 additional employees who get their stores from the Subsistence Department.

The Committee thinks it important that fresh vegetables be furnished employees, and if necessary, believes that the government should undertake truck-gardening for this purpose.

Some complaint is made as to the cooking and the preparation of the food, which the Committee believes is well founded, and thinks increased attention should be given to this matter, as well as to the table service at the hotels.

Good Housing

The Committee on housing employees of the Canal has only words of commendation. It gives as information to the Committee of the Whole [a] description of the different houses which the government has adopted as standard. These houses are simply but comfortably furnished. . . .

At each important town a station hotel, like the one at which the party lunched on March 1st, has been erected. Dining room 30 × 100 feet. Second story: sixteen large rooms 16 × 16 feet, eight toilets. In another building, connected by piazza, is assembly hall, billiard and pool room, and above dormitory of sixteen small bed rooms 10 × 12 with showers and toilets.

There were over 2,100 dwellings of different sizes which came into possession of the Commission from the French. The poorest of these were destroyed and the balance have been repaired and put in first-class condition. The government is adding constantly to the number of dwellings, and is constructing many at the present time.

The excellence of the housing seems to us one of the striking features of the Canal work. Those familiar with labor quarters in any section of our own or other countries cannot but feel that the housing offered on the Isthmus is the best in the world. . . .

> Those familiar with labor quarters . . . cannot but feel that the housing offered on the Isthmus is the best in the world.

Pay Scale for Laborers

Mr. Farwell, Chairman of the group on Efficiency of Labor, was presiding over the meeting. He called upon Mr. Worthington, of Cincinnati, to read the report of that group. The report was as follows:

The labor can be most conveniently classified for our purposes as skilled and unskilled.

Before our inquiries upon the ground we had supposed the supply of unskilled labor presented a problem of continuing difficulty, but we were assured on all sides that this problem was practically solved.

The unskilled labor now employed consists in the main of West Indian negroes, Italians and Spaniards. A considerable body of Greeks is expected shortly. These white men are sending home for their families and friends, which is encouraging as showing their liking for the work.

The negroes are paid thirteen and one-third cents silver per hour, and their meals; the white men, forty cents silver per hour, and they furnish their own meals. Every laborer of each class is furnished lodging, light, fuel, water and hospital service free of charge.

Nine hours constitute a day's work, viz: from 6:30 to 11:30 A.M. and from 1 to 5 P.M.

The difference above mentioned in the scale of wages indicates a marked difference in the efficiency of

The Workers in Panama Were Well-Paid

They are said to be a contented lot, these thirty-five thousand men who are doing this wonderful thing. There is a certain fascination about the "big job," of course, and nowhere on earth is there a body of employees so well cared for as this one. Besides receiving higher wages for the same kind of work, they are obliged to spend less for their living than would be the case anywhere else on the globe. Their lodgings are free and their food is furnished them at very low rates. A West Indian negro, for instance, who, on his native island, might earn a quarter a day for a part of the year, here receives a dollar a day, free lodgings and a day's food for thirty cents. Among the white Americans, a married man is given a furnished house, electric lights, kitchen fuel and moving expenses free, and his food, ice and clothing at reduced prices. Medical and hospital service cost nothing. Nowhere else could he live so well and save so much. In addition to all this, he is given a month off every year with pay and special traveling rates. We became acquainted with one young couple who had been there six years and they told us that they expected to go back to the States after a while with enough money saved up to render them almost independent.

SOURCE. *Mary L. McCarty,* Glimpses of Panama and of the Canal, *Kansas City, MO: Tiernan-Dart, 1913. pp. 148–150. www.openlibrary.org.*

the work; and if the comparative results achieved were correctly stated to us, the negro is the better paid of the two. For, reducing these figures to a common basis by taking gold to silver at two to one, according to Isthmian

Canal construction workers enjoy some time off at the Miraflores Lock. (© Bettmann/Corbis/AP Images.)

usage, and counting the white man's meals at forty cents gold per day, at which rate he can procure them from the Isthmian Canal Commission, he receives net per day $1.40 gold, while the negro receives sixty cents gold, thus making a ratio of two and one-third to one, while the comparative ratio of efficiency in work as given us is approximately three to one.

The reasons assigned for the greater efficiency of the white laborer are greater intelligence, physical strength, endurance and vitality. When instructed as to a new piece of routine work the lesson lasts and needs no repetition, while the negro needs constant supervision. The white man attacks his work more vigorously and he keeps in good health. This matter of health has led to requiring the negro to eat the food supplied by the

Canal Commission as, when left to his own desires, he preferred the toothsome but innutritious yam to the strength-giving bean. As a result, the efficiency of the negro has greatly increased and [he has] worked more steadily. The President [Theodore Roosevelt] having recommended the change, it is interesting to note what has been accomplished.

No trouble has been experienced lately in getting an adequate supply of each class of labor, and none is now anticipated. Of course the force must be constantly recruited to keep up with the present demands as well as to be prepared for the future.

It is estimated that over eighty per cent of the black force has changed during the year; and yet, while over ninety-five per cent of the whites in a given force can be relied on for constant service, less than seventy per cent of the negroes will be available.

We were assured that the force now on hand was as large as could be profitably employed with the present equipment. An increase in cars and tracks will open the field for more men; but there seems no doubt that men, white and black, can be obtained in sufficient numbers as needed. We say, "white and black" for two reasons:

First. Because we were assured that notwithstanding the superiority of white labor by every test, it is desirable to employ both kinds to prevent each from fancying it could control the situation.

Second. Because Chinese labor is not wanted. While the Chinese coolie would meet every requirement of intelligence, strength and vitality, yet his introduction would lead to difficulties of administration and segregation which it seems desirable to avoid unless the confronting of the subject becomes necessary.

The only serious trouble now noticed is the difficulty of securing efficient gang bosses from the ranks, the general scale of intelligence not being quite equal to this demand.

Compensation for Skilled Workers

Passing now to the skilled labor: this force has been recruited almost entirely from the United States. The method now employed to secure it is by employment agencies at home. Inquiry is there made as to the qualifications of applicants, and on arrival the men are assigned to the class of work with which they are familiar. Naturally the scale of wages, which includes free lodging, light, fuel, water and hospital service, is higher than is paid for similar work at home.

"The scale of wages, which includes free lodging, light, fuel, water and hospital service, is higher than is paid for similar work at home."

Eight hours constitute a day's work: from 7 to 11 A.M. and from 1 to 5 P.M., and each man is entitled to forty-two days vacation, and if he needs it, to thirty days sick leave, on full pay.

The inducements offered have been sufficient to tempt an adequate supply of competent men. Of course some come who are unfitted for the work, and others who, though fit, have not the adaptability to meet the surroundings. These are dismissed as soon as their deficiency is discovered. But the good man stays, is pleased with his work, and makes more money than he could at home.

As vacancies occur in the higher ranks they are filled, if practicable, from below, and some of the men whom we met have reached their present positions by several promotions from inferior positions in which they had proved their capability.

The chiefs with whom we came in contact impressed us as a highly intelligent, picked force—such a representation as might be expected from pursuing the principle of the survival of the fittest.

We made inquiry as to the efficiency of the labor, both skilled and unskilled, as affected by climate, and as to the difference in the product of a day's labor on the Isthmus, compared with that of the same type of man at home, and were told there was practically no difference aside from the allowance already mentioned for vacation and sick leave. That this will continue indefinitely for the whole body of skilled labor may be doubted, as eighteen months or two years seems too short a period from which to draw such a conclusion. But past experience proves at least that the tropical climate has not the enervating influence upon the efficiency of the healthy man interested in his work that had been feared.

As to the unskilled labor there need be no apprehension, for the climatic conditions are those to which the men of that class are accustomed.

We asked also as to the difficulties in handling labor from insubordination, strikes and other troubles, and were advised that none had been experienced or was apprehended. We heard of but one organized labor union, that of the steam shovellers, consisting of 180 men.

The situation is so different here from that in the normal labor market that the occasion for such troubles is not likely to arise. There is but one employer and house owner, and he cannot be expected to furnish lodgings to those who do not work. Those who quit work will be compelled by force of circumstances to go elsewhere, and cannot stay and intimidate those who wish to continue.

Praise for Workers' Zeal

In conclusion we wish to call attention to the sympathy with the work and the zeal and intelligence of the men with whom we came directly in contact. They were young men and picked men. One and all, they spoke with unhesitating faith in commendation of the plan of the work, of the ability of the Chief Engineer [John F. Stevens], and of his fidelity and zeal, and of the practicability of completing the Canal upon the lines now proposed.

And they bore witness to the existence of a like faith in those under them, and of competition among them as to who should make the best showing for his day's work, if an artisan, or for his squad or department, if of higher grade.

> [The men in charge] have created an *esprit de corps* which has permeated the whole body of men under them.

We believe that the men in charge have solved the labor problem, and we have only words of praise for what they have accomplished. They have created an *esprit de corps* which has permeated the whole body of men under them, and will lead to a zealous prosecution of the work. It is of the utmost importance that the spirit thus inculcated should be fostered and encouraged by all in whose hands the execution of this stupendous task will fall.

VIEWPOINT 6

Black Panama Canal Construction Workers Are Not Treated Fairly

David Newton E. Campbell

David Newton E. Campbell, a successful American doctor and author born in the British West Indies, took a vacation trip to Panama in 1908. In his book about the visit to the canal he devoted considerable attention to racial prejudice in the Canal Zone. In the following excerpt he describes his interview with a black West Indian laborer who told him about the low pay, poor living conditions, and general mistreatment of the black canal workers. They were forced to buy their meals, he says, although the food was not to their liking and sometimes there was not enough to go around. They received no respect from the authorities. And if they were injured on the job, as many were, they received no compensation and could not be rehired unless they signed papers agreeing not to demand any. The author also reports that he was himself affected by racial prejudice in the canal area despite his US citizenship.

SOURCE. David Newton E. Campbell, *Searchlight on the Panama Canal; or, America's Greatest Enterprise*. Baltimore: Baltimore Book Company, 1909, pp. 59–64. Courtesy of Baltimore Book Co.

"Are you laborers well paid for your work?" I asked an intelligent Barbadian with whom I talked at Gatun [Panama].

"No sir," he said emphatically, "most of us were told at home that we would get on the Isthmus 'plenty work and good pay,' but we have been sadly disappointed. An Isthmian Canal agent told us at home that able-bodied men as I am could get from two to five dollars per day working at the canal. But I have not received a dollar per day yet since I came to this Isthmus, nearly eighteen months ago.

> While the wages [are] small a fellow might try to get along, if these bosses would treat a man as if he were a living being.

"But while the wages is small a fellow might try to get along, if these bosses would treat a man as if he were a living being. But instead you are 'cursed' at and treated like beasts all the time. We poor West Indians have got it to the very limit of endurance. In that haughty and domineering way, you are ordered, sometimes knocked down, kicked, and if your manhood resents the inhuman treatment given to you, then you are up against it like a ship tossed by an angry sea.

"I have known friends of mine who were beaten on the train and pushed off after paying his fare. But what can one do in such instances? You have practically no redress. Is it that way the white people treat colored people in America? Heavens! I don't know how they could have endured it. Certainly, I am trying to bear it for the time being, but God knows that as soon as I can save enough to pay my way home, this place will never see my face, no sir, never! never!"

"No! in America the colored people are treated first-class. Have you not read that in America we have more wealthy and great negroes than in any other country in the world? Why, to be sure. But they have the Constitution to back them up. Certainly, I have heard that the colored people in the far Southern States have a hard time sometimes, but that spirit of hate and bitterness is yearly dying out. I guess you colored people from the West Indies do not understand the Americans, but they are fine people—the most liberal in the world," I replied.

"Well, I won't contradict you, sir, but from what I have seen of those with whom we have to contend here on the Isthmus, I—I—I, well—well. Probably they pick these out and send them down here purposely. There are a few fine men among them, but that class can be counted upon the fingers of one hand. And with all the bad treatment, they don't expect you to say a word; if you do, then you run a big risk of being reported and then arrested for something. And then, heaven smile on you, for when a complaint is lodged against a colored man here, nothing saves him from jail, and jail sentence here! ah! ah! when a poor negro serves three months here in jail, I believe he should be immuned from any future punishment, either now or after death, however black his sins might be.

> When a complaint is lodged against a colored man here, nothing saves him from jail.

Low Wages for Blacks

"Well, I know each month millions are spent for canal work, and a big payroll is audited, but the money does not come our way, because we get only 20 cents balboa an hour for labor or ten cents American money. And we work only eight hours per day, which makes the earnings of a laborer only 80 cents per day. Out of that amount the Commission takes 30 cents per day for mess, as they compel us to eat at the mess kitchen, and whether you eat or do not, the 30 cents are deducted just the same.

"If the laborer is a married man, then he gets his full pay and nothing is deducted from his wages. You can see that at ten cents an hour of eight hours daily that's

Racist Attitudes Led to the Distinction Between "Gold" and "Silver" Canal Workers

Although white workers on the Panama Canal were well paid and provided with many benefits, black workers were not. Racism, although not acknowledged to be US government policy, determined the organization of the project. The labor force was divided into "gold" and "silver" employees. Theoretically, this was because skilled workmen were paid in gold-backed US currency and unskilled laborers were paid in silver coin. Because, after the first hires, no skilled blacks—or any American blacks for that matter—were hired, and unskilled white men were discouraged from applying, this amounted to racial segregation. William R. Scott, in his 1912 book *The Americans in Panama*, wrote: "The idea was adopted as the best means for the Government to draw the color line—a practice it would not attempt under the Constitution in the United States." As time went on, the division became more strict, and with few exceptions any skilled blacks on the Gold Roll were moved to the Silver Roll.

Many distinctions were made between gold and silver employees. For example, the quality of silver employees' housing and meals was much lower. Although it was reasonable to pay unskilled workers less than skilled ones, there was prejudice in hiring, and black laborers were looked down upon more because of their color than because of their work performance. The vast majority of canal workers were West Indian blacks who, being non-Americans, were termed "aliens." They did not have the same privileges as the Americans and were often mistreated. American blacks were also subject to mistreatment, as described by David Newton E. Campbell in his book *Searchlight on the Panama Canal, or, America's Greatest Enterprise.* Although a well-educated doctor and a US citizen, Campbell was forced to pick up his mail at the "alien" post office in Panama. There he endured long lines even when there was no waiting at the window for white visitors.

Few writers of the era mentioned these problems, and the American public would not have seen anything wrong or unusual about segregation on the basis of color. The Supreme Court had recently ruled that even in the United States segregation was not unconstitutional as long as facilities for the races were equal. In Panama the facilities were far from equal, but because this was attributed to level of skill, it remained unchallenged.

80 cents, and six working days will give a laborer \$4.80 per week. At 30 cents per day for seven eating days that's \$2.10 for mess per week. Hence, after deducting our mess bill we have coming to us only \$2.70 per week, and at the end of a month of four weeks we get \$10.80. But that is when we work all along. And many times we are laid off for several days, no work at all, so that you can't always depend on the \$10.80 per month. Out of this amount we have to pay for clothes washing, buy shoes, clothing, and meet our other incidentals. So you see, it is so arranged that we cannot make much. It is a fact that sometimes we make extras, but that does not amount to very much, and it is not often that extras come our way."

"But has not the Commission just cause for compelling laborers to eat at the mess kitchen?" I asked.

"Perhaps it is just from their point of view. They claim that the laborers will not feed themselves well enough to do the canal work unless fed with American proteids; but every reasonable mind knows that each question has two sides. The Commissioners have printed their side of it in the Canal Record and other papers. But has anybody heard the poor laborers' side of it? Has anything been allowed to go in print from the second party in this unjust transaction? No! no! Not a single word!

"Every one knows or ought to know that West Indians have not been accustomed to feed on American food; their food is not our food; their manner of making bread is not our manner; taste is not our taste; yet, they have undertaken to regulate our palate and forcibly compel us to accept whatever they feel like giving us to eat. It is true that they give us a West Indian to cook the food for us, but what is that? I tell you, sir, it is a pretty serious thing when a man is placed in the position where somebody else has the say-so for what you shall eat without any alternative. But we have no voice pro or con in the matter of what we eat. . . .

Not Enough Food

"Sir," he continued, "when the order was first issued that all unmarried laborers will be compelled to eat at the mess kitchen, I was at Cristobal and saw men fall in line with their dinner pails, and pass up in order to the cook, with the police standing there to see that discipline was maintained. As the crowd marched up the cook ladeled out the food and each passed on. Presently the cook says all the food prepared is served and he hadn't anything more. There were men in the line who did not receive a morsel of anything to eat. Well, those poor fellows had either to remain hungry or try to get a mouthful from the more fortunate ones who had something to eat.

> Whether you get the food or the cook 'runs short,' when pay day comes around, your 30 cents a day is deducted.

"When the food is served, as there is no table provided for West Indians as is done for European laborers, then they scatter about and sit on the grass or on a log or lean against a fence, or crouch under a shed. But whether you get the food or the cook 'runs short,' when pay day comes around, your 30 cents a day is deducted like anyone else. Well, as is natural to expect, many of the laborers refused to accept such treatment. That class has no alternative but to 'quit.' Because if you do not eat at the mess kitchen, you are not allowed to sleep in the camp. Yes, if you refuse to eat what they give you, then you must take bag and baggage. 'No eat, no sleep.' Then, remember, every house belongs to the I.C. Commission, small private houses cost a fortune. Small rooms in which a person can scarcely stretch himself cost from 20 to 30 dollars per month. . . .

No Compensation for Injury

European laborers are very much better provided for than West Indians. They get from 30 to 40 cents per hour

The SS *Ancon* brings 1,500 laborers from Barbados to Panama. Reports from the time show that black workers on the canal were treated poorly compared to their white counterparts. (© Time & Life Pictures/ Getty Images.)

for their labor, doing the same work as we do. They also have tables on which to place their food and benches or chairs on which to sit while eating.

"Then again, if a West Indian laborer gets hurt during the performance of his duty, and his injury is sufficiently serious, he is sent to the hospital, where he remains until he is better. During his stay in the hospital he is not paid a single cent. When he is well again he is made to sign a release or exemption paper. It does not matter how many children the injured laborer has, whether his wife is able to take care of the children or not. This paper of release is to exonerate the Commission or Panama Railroad Company from all blame for the injury the West Indian received during the performance of his duty. If the unfortunate victim loses his hand or foot in the accident, when he signs the release, he is paid one dollar gold, and that's all he gets. But by signing the release, he becomes eligible to get his job again, or some other job under the I.C.C. or P.R.R. If, however, he refuses to sign the paper of release when presented to him, then he gets neither the dollar nor his job. There would be no work for him again. . . .

"And I know several instances where people were killed by the train, and the Panama Railroad gave the coffins in which they were buried; in other instances they would give not a single thing, neither coffin nor any money, and friends of the unfortunate victim have to go around and take up collection to bury the dead if he was not a member of some secret society." He then told me the names and addresses of three victims of accidents who lost limbs and had to sign a release before they could be re-employed.

> There is a deep-seated and wide-spread *'dislike'* if not *'hate'* in the bosom of the natives for Americans, due principally to color prejudice.

"I saw some time ago that a Liability Act was passed in Congress to compensate employes who were injured in the performance of their duties. I think special reference is made in it for employes of the Panama Canal," I said.

"Well, sir, that might be so, but it might be intended for Americans only." . . .

"Poor colored laborers can't form any union here. Most of the laborers are here on contract, and they have to do what they are told and accept just what they can get. But it is not the colored laborer alone who feels the sting of American prejudice on the Zone. . . .

The author, who spent nearly a year studying the conditions on the Canal Zone and adjoining cities, will say: there is a deep-seated and wide-spread "*dislike*" if not "*hate*" in the bosom of the natives for Americans, due principally to color prejudice. That is the pivotal point. Nearly all the Panamanian policemen are colored, and probably ninety per cent of the population. And . . . "clashes" [with American workers] are sure to occur until race hate subsides.

VIEWPOINT 7

Exemption of US Coastwise Shipping from Canal Tolls Is Not Fair to Great Britain

Henry Erle Richards

When the Panama Canal was nearly completed and plans were made for its operation, the US Congress passed a law exempting American "coastwise" ships (those that traveled only between US ports without visiting other nations) from the tolls charged to other ships. This resulted in controversy. Great Britain, in particular, complained that it was unfair and that it violated a 1901 treaty that said ships of all nations were to be treated equally with respect to their use of the canal. In the following viewpoint Henry Erle Richards, a professor of international law and diplomacy at All Souls College (a part of England's Oxford University), explains

---

SOURCE. Henry Erle Richards, *The Panama Canal Controversy: A Lecture Delivered Before the University of Oxford on October 25, 1913.* London: The Clarendon Press, 1913, pp. 3–4, 7, 22–34. Courtesy of The Clarendon Press.

why he believes the treaty forbids such an arrangement and why exemption of US ships from tolls would be harmful to Great Britain.

The near approach of the time when the Panama Canal is to be opened for public traffic has brought to a head the controversy between this country [Great Britain] and the United States in regard to the tolls to be imposed on British Shipping which makes use of the new waterway. There are rumours that the President of the United States intends to propose the repeal or suspension of the clauses of the Canal Act to which Great Britain has taken exception. . . .

The United States claim that they have the right to discriminate between their own vessels and those of this country, or, in other words, that they can remit tolls on American vessels without giving corresponding relief to British ships, or can charge lower tolls on their own vessels as compared with those imposed on the ships which fly the British flag. . . . His Majesty's Government, on the other hand, contend that the United States have bound themselves by the Treaty between the two nations of 1901, generally known as the Hay-Pauncefote Treaty, to make no discrimination as against British vessels, and that the claim made by the United States and the powers conferred by the Canal Act are contrary to that Treaty in so far as they affect or may affect the shipping of this country. . . .

The advantages of some waterway for sea-going ships from the Atlantic to the Pacific has been recognized from the earliest times, indeed it is but necessary to glance at an atlas to realize the immense commercial importance of some direct connexion between the two oceans. Here is this great continent stretching across the face of the globe from North to South forming an impenetrable barrier to the traffic of the world as it goes from East to West or from West to East. . . .

In 1898 the necessity for some short cut by sea between the two oceans was brought home to the people of the United States in a practical way. They were at war with Spain, but the [US battleship] *Oregon* on the Pacific Coast could not be brought to the area of operations in the Atlantic except by travelling round Cape Horn, a voyage of some 13,000 miles; an impressive illustration of the fact that unless there were some Canal, the United States fleet on the two seaboards could not be united except after a delay which might, in some circumstances, prove fatal. Moreover, new territories in the Pacific had been just acquired, the Hawaiian Islands and the Philippines; and for them the Canal route was specially needed. For these reasons it was realized to be essential for the United States to have some Canal, and to have that Canal under United States control. But, before this could be done, it was necessary to get rid of the restrictions imposed by the Treaty of 1850. . . .

The Hay-Pauncefote Treaty of 1901

There is no word in the new Treaty nor is there any suggestion to be found in the negotiations that the parties intended to abandon the "great design" of the Convention of 1850, namely, that the Canal was to be "neutralized," in the sense that it was to be open for the benefit of the two nations, and of all other nations who would agree to treat it from that point of view; and that it was to be for this common use on equal terms. If that fundamental condition was to be abolished, and the whole character of the enterprise to be altered, the change must have been the subject of discussion, and the Treaty would have given effect to it expressly. . . . It is very difficult to believe that the parties could have intended to make such a change. Here was a bargain made in 1850 between commercial rivals to ensure equal treatment for their trade; that bargain was in force and bound both nations; it is, to say the least of it, extremely unlikely that Great

HOW THE PANAMA CANAL SHORTENED SHIPPING ROUTES

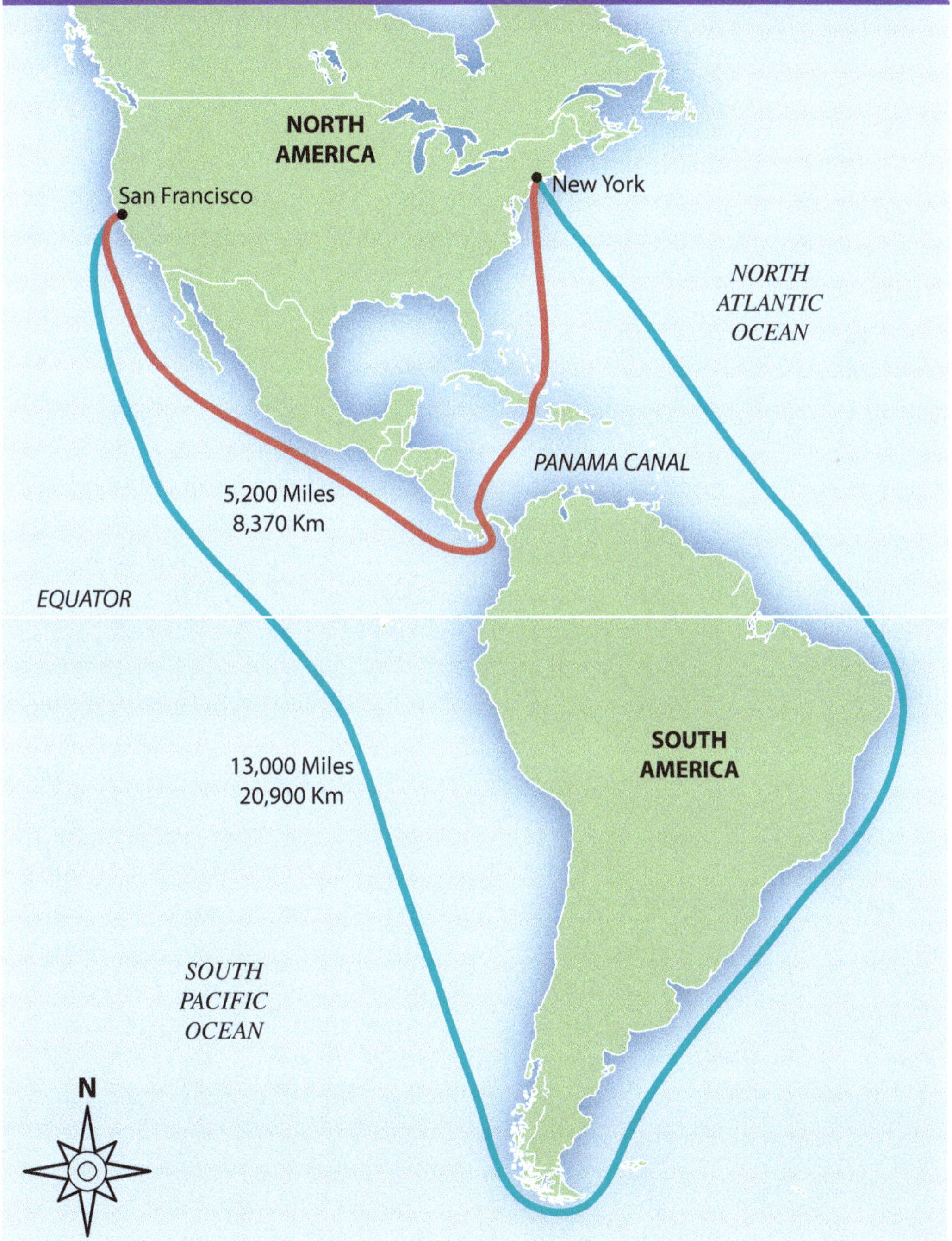

Britain intended to make a present to her rival, not only of the right to control the Canal, but also of the right to discriminate against her trade. There would have been nothing to be gained on the British side by such a surrender of her rights under the existing Treaty, indeed, if the United States were to be entitled to the discrimination that is now claimed, there would have been but little use in having any Treaty at all.

But the intention of the parties can be placed beyond doubt. It can be shown by direct evidence that it was the purpose of those who actually took part in the negotiation of the new Treaty to preserve the condition of equal treatment. . . .

Then we come to the direct provision contained in the first Rule of Article III, that the Canal is to be free and open to the vessels of all nations on terms of entire equality with no discrimination. The words are not ambiguous or vague, there is no exception or qualification to them, and there seems no reason why they should not be construed in their ordinary sense. If it had been intended to abandon the principle of equality which had governed the relations between the two countries from the outset, one would have expected to find at this place some express exemption of the vessels of the United States. But there is none. . . .

> [President] Taft argues that the Treaty imposes no fetter on the sovereign rights of the United States . . . in the matter of tolls.

The American case is sufficiently expounded in Mr. [President William Howard] Taft's Memorandum to enable us to appreciate the main lines on which it is based. The effect of it is that there is no restriction on the United States in the matter of tolls save this—that they must give to British vessels the most favoured nation treatment, or, in other words, that they cannot charge higher tolls on British vessels than those charged on the vessels of any other foreign

Power; but that as between British vessels and United States vessels they can discriminate to any extent they please. Mr. Taft argues that the Treaty imposes no fetter on the sovereign rights of the United States to do what it pleases in the matter of tolls, because, as he says, the rules in Article III are a mere declaration of policy on the part of the United States in order to effect the neutralization of the Canal in time of war; in other words, that all nations who agree to treat the Canal as neutral in war are to have like treatment, but that the power of the United States with regard to their own vessels is not covered by or referred to in the Article. In the view thus presented Article III refers only to the action of foreign Powers, and the words "all nations" in Rule I refer to foreign Powers only and do not include the United States. . . .

The contention of the United States, so far as it has been yet developed, is not a sound contention; it is a contention based on too narrow a reading of the words of the Treaty, and can be maintained only if previous history and the negotiation between the parties are put out of sight. Diplomatic agreements must not be interpreted by the same strict rules of construction as are applied to legal documents; it is the intention of the parties which must govern treaties rather than the particular form in which effect is endeavoured to be given to that intention; and in this case, as I submit to you, the intention of both parties has been throughout manifest beyond doubt. . . .

The Tolls to Be Imposed

The Panama Canal was pushed on, and the project has been carried through to a successful issue with that splendid energy and fertility of resource which is characteristic of the American nation. And now the work is approaching completion, and the question of construction of the Treaty has been raised in a direct way by a publication of the list of tolls to be charged on shipping using the Canal. That list has been issued under the authority of an

A US steamship, the SS *Ancon*, officially opens the Panama Canal on August 15, 1914. (© Bettmann/Corbis/AP Images.)

Act passed in 1912 entitled the Panama Canal Act, under which the Legislature of the United States have assumed powers with regard to tolls, which, in the view of His Majesty's Government, are contrary to the stipulations of 1901. . . . In the first place, it provides that the tolls when based on net registered tonnage shall not exceed one dollar and twenty-five cents per ton, nor be less, *other than for vessels of the United States and its citizens*, than the estimated proportionate cost of the actual maintenance and operation of the Canal. In the second place, it enacts positively that no tolls shall be levied upon vessels engaged in the coastwise trade of the United States. . . .

Coastwise shipping, according to the law of the United States, is restricted to vessels flying the United States flag, and it is contended that since foreign vessels cannot take part in coastwise trade, there is no competition in regard to that trade between United States vessels and foreign vessels, and therefore no discrimination within the meaning of the Treaty. But if the basis on which this argument rests be examined it will be found to be wrong in fact. Coastwise vessels, as defined by the law of the United States, will compete with foreign-going vessels, and a discrimination in their favour will defeat the protection which Great Britain alleges the Treaty has given to her shipping. "Coastwise" is not a term of art, it has no precise limits, its construction and extent depend on the municipal law of each country and vary according to that law. In England coastwise trade is open to the ships of all flags, but in the United States and many other countries it is reserved for the ships of nationals. But the extent of the reservations is not uniform. It may be limited to voyages from port to port which do not involve the crossing of the open sea or the passage through the territorial waters of some other State; on the other hand, it may be extended to cover distant voyages across open seas with calls at foreign ports. . . . Under [US law] trade between any port of the United States and the Philippines or the Hawaiian Islands or Porto Rico [Puerto Rico] is coasting trade. It follows that American vessels trading between New York and the ports of Washington State on the Pacific Coast, or between New York and Manila, to take two instances, are all to be within the tolls exemption. It is manifest that trade of so wide an extent as this must compete with foreign-going ships. For instance, take the case of trade to the Western coast of the United States. British ships going from London to San Francisco through the Canal must pay tolls, and those tolls must increase the cost of the goods delivered at San Francisco, and thereby make them the less able to compete with goods from New York to the same port. . . .

> British ships going . . . through the Canal must pay tolls, and those tolls . . . make them the less able to compete.

It might be possible to defend an exemption limited to coasting trade, if that trade were confined to the small craft which ply between neighbouring harbours, carrying goods of small value and purely of domestic trade. . . . But coastwise trade as defined by the law of the United States is "coastwise" not in fact, but by a fiction of the law . . . in fact, it is not coasting trade at all, it is not trade of purely domestic concern, and it does compete with foreign trade. Therefore, if this exemption from tolls were conferred, United States vessels would be at an advantage as compared with other vessels engaged in the same trade by reason of a discrimination in breach of the Treaty.

Exemption of US Coastwise Shipping from Canal Tolls Is Warranted

Jahrbuch des Völkerrechts (Yearbook of International Law)

In the following excerpt from a translated German publication, the author argues that Great Britain's interpretation of the Hay-Pauncefote treaty, which it claims the United States has violated, is mistaken. No one ever thought the United States would pay for the canal without expecting to control it and receive a return on its investment, he says. Thus the treaty was intended to place "all nations" on equal terms with each other in their use of the canal, not on equal terms with the United States itself. Moreover, the inclusion of the United States among "all nations" as referred to in the treaty does not make sense, since it contains provisions concerning wartime that by their nature could not apply to the canal's owner. The United States could not defend the canal if it

---

SOURCE. Translated in George C. Butte, *Great Britain and the Panama Canal: A Story of the Tolls Question*. Washington, DC: United States Government Printing Office, 1913, pp. 1–3, 66–73. Courtesy of the United States Government Printing Office.

had to follow rules prohibiting belligerent actions, for instance. In his opinion the only reasonable conclusion is that the treaty does not restrict the United States' power to make any laws it sees fit about the canal and the charges for its use.

He deceives himself grievously who believes the United States made the stupendous sacrifice of human energy and public money necessary to build the Panama Canal, "the greatest liberty Man has ever taken with Nature," with any other purpose in view than the national advantage of the United States—commercial and, above all, political advantage. A critical study of the diplomatic history of the various Isthmian Canal projects from 1835 . . . down to the present time will create no doubt in the mind of any except the most naive that the United States ever had any other intention. Equally clear also is the fact that the United States never thought of selfishly appropriating the canal as a national monopoly. It was to be a great open highway to bring remote nations and peoples into closer relation with one another; it should be dedicated to the commerce of the world as an exalted agency of peace, in the control of a liberty-loving people, for the promotion of human happiness and the spread of the blessings of civilization. Nor is there anything necessarily incompatible in these positions. No one underestimates the benefits to mankind of the Suez Canal because the builders thereof draw a 30% dividend annually.

> No sane statesman . . . believed the United States would undertake to build the canal under any other than practically sovereign control.

For the past eight years the American nation had given the best of its brawn and brains and offered up its millions of treasure to the execution of the great plan, that had occupied the dreams of her earliest statesmen;

and the people were rejoicing over the dispatch of the work and the prospect of the early realization of the national benefits they expected to receive as a return upon the enormous investment. Accordingly the Congress of the United States, in anticipation of the opening of the canal in the course of the year 1913 and in order to give the shipping interests of the world sufficient time in which to adjust themselves to the far reaching changes in the course of navigation likely to follow the opening of the new waterway, enacted a law entitled "An act to provide for the opening, maintenance, protection, and operation of the Panama Canal and the sanitation and government of the Canal Zone." It was approved by President [William Howard] Taft on August 24, 1912, with the comment that he regarded the bill on the whole "to be one of the most beneficial that has passed this or any other Congress." . . .

> The parties never intended that vessels . . . should be entitled to use the canal 'on terms of entire equality'—with the United States.

The Intent of the Hay-Pauncefote Treaty

The Hay-Pauncefote treaty . . . itself contemplates a condition of practically sovereign rights. No sane statesman, knowing the strategic value of the canal and the enormous sacrifices to be made, then [at the time the treaty was signed] believed the United States would undertake to build the canal under any other than practically sovereign control. Article II grants that the United States "subject to the provisions of the present treaty, shall have and enjoy all the rights incident to such construction as well as the exclusive right of providing for the regulation and management of the canal." What larger rights has the United States since acquired? The rights which she acquired by treaty with Panama in 1903 are indeed expressly made subject to the restrictions of the Hay-Pauncefote treaty.

The conclusion is irresistible that the parties never intended that "vessels of commerce and of war of all nations observing these rules" should be entitled to use the canal "on terms of entire equality"—with the United States; but on terms of equality among themselves, so that there should be no discrimination against any of them, giving nothing—absolutely nothing in return for the pledge of the United States to treat them all fairly and equitably at the canal.

The idea that the United States is to be included in "all nations observing these rules" leads to other equally untenable and preposterous conclusions. The United States alone "adopts" the rules for the neutralization of the canal; and the treaty contemplates that these rules may be amplified and supplemented, and a superstructure built thereon (for they are only a "basis") by the United States and by it alone, as it may deem proper or necessary to make the observance of the neutrality of the canal more certain. It is possible, for instance, for the United States to lay down the rule that no warship cleared for action shall be permitted to enter the canal, or that no warship shall be permitted to enter whose commander will not agree to be the guest of the United States during the passage—rules that can, in their nature, be observed only by other nations. The controlling principle is that the treaty recognizes the power of the owner of the canal to prescribe the conditions upon which others, non-owners, may use its property. . . . This power extends not only to the supplemental rules which it may hereafter adopt but also to those which it has already adopted, as declared in the Hay-Pauncefote treaty.

To say that the United States is included in "all nations observing these Rules" is to contend that her right to use the canal "on terms of entire equality" (with the United States!) is conditioned upon her demonstrating (to the United States!) that she is, as a belligerent, respecting the neutrality of the canal (the property of

US president William Howard Taft (center) visits a Panama Canal lock site in 1910. (© Time & Life Pictures/ Getty Images.)

the United States!) in conformity with the rules adopted by the United States! This contention is simply incomprehensible.

Nothing was farther from the minds of the parties than that the United States, as the nation solely charged with enforcing the neutrality of the canal, should be prohibited from taking any belligerent action it saw fit to protect the canal. . . .

It Is Not Logical to Believe That the Treaty Prohibits Toll Discrimination

If we are prohibited [by the treaty's provisions referring to "all nations"] from discriminating as to tolls, we are also, of necessity, prohibited . . . from blockading the canal or exercising any right of war or any act of hostility in the canal; and this is technically true, although our enemy, in case of war, might seek to fight us there and even to destroy the canal.

But it is still further provided in this same article that "vessels of war of a belligerent shall not revictual nor take any stores in the canal except so far as may be strictly necessary." If we should get into a war with Japan or Germany or any other nation except Great Britain, we would find ourselves prohibited . . . from taking on food supplies or stores of any kind in the canal.

It is also further provided in this same article that "no belligerent shall embark or disembark troops, munitions of war, or warlike materials in the canal." This would prohibit us from supplying the fortifications which we are now placing there with ammunition and guns to be used in the protection of our property against a nation, other than Great Britain, with whom we might engage in war.

Carried to its logical and ultimate

Need to Defend the Canal Gives the United States Special Status

Whenever, and as long as the United States deems it necessary for the protection of the canal and the enforcement of its rules, it may, in contravention of rule 1 of Article III, close the canal partially or entirely even against nations "observing these rules"; and it may, for the same reasons, commit any of the warlike acts prohibited in rules 2–6 inclusive. If the United States is bound to "observe these rules" how could it either enforce its rules or defend the canal against attack, or assert the exclusive rights of control and management granted to it by Article II of the treaty? These things seem so obvious that it is astonishing that anyone would seriously contend to the contrary. That the words "all nations observing these rules," from the very nature of things, did not include result, a violation by us of any of these prohibitions or restrictions would forfeit our right to use the canal on any terms whatever. The result would be that at a cost of more than four hundred millions of dollars, without any help from anybody, and without any co-partnership with anybody, we have constructed this great work and bound ourselves to forever maintain and operate it, at our sole expense, without any right or power to show the slightest favor to any of our vessels of war or commerce, either coast-wise or of foreign commerce. . . .

In all this I am speaking of the question of "power" and not the question of "policy." It may be wise for us, as a mere matter of policy, to practice such altruism. But the treaty does not require it, and it would be most unwise to concede it.

SOURCE. *Speech of Hon. J.B. Foraker Before the Richland County Lincoln Association at Mansfield, Ohio, February 12, 1914. www.openlibrary.org.*

the United States was evidenty so clear to the parties in 1901, that it never occurred to them necessary to say "all *other* nations observing these rules," as the British side now contend would have been necessary. Indeed, the latter form of expression might justify a doubt whether the United States was not, too, bound to observe the rules, which is precisely what was not intended. . . .

It will be remembered Secretary [of State John] Hay declined to accept the form, "all nations which shall agree to observe these Rules" because of the "strong objection entertained to inviting other powers to become contract parties to a treaty affecting the canal"; clearly indicating that the expression was regarded as relating to other nations than the United States. Other nations were to agree with the United States, under whose exclusive control the canal was to be built, to observe the rules adopted by the United States. In no other way, in fact, was an agreement conceivable. When this idea of agreement to observe the rules was eliminated and the fact of observance was substituted therefor in the amendments "all nations observing these rules" and "such nations", two things were accomplished 1) Great Britain was placed on the same footing as other nations and 2) the objection of the United States was met against conferring upon *other nations* a contractual right to use the canal. So thought Lord [Henry Petty-Fitzmaurice] Lansdowne and so Secretary Hay. No better evidence is to be had, that the provision "the canal shall be free and open to the vessels of commerce and of war of *all nations observing these rules* on terms of entire equality," does not embrace or relate to the United States; and that, therefore, neither rule 1, nor any of the other rules expresses any intention to restrict the powers of the United States as to granting a subsidy equal to the tolls, or refunding the tolls (as other nations may do), or as to granting a direct exemption, which is essentially the same thing; or as to imposing on her own ships exceptional burdens and onerous conditions not exacted of other nations—in a word, that the treaty was never intended expressly to restrict the power of the United States either in time of war or in time of peace to legislate as to her own vessels of war or of commerce, coastwise or foreign, in any way she sees fit. This is a sane and reasonable conclusion and the only one that harmonizes with all the provisions of the treaty.

VIEWPOINT 9

Repeal of the Law Exempting US Coastwise Shipping from Canal Tolls Was Morally Right

Joseph P. Tumulty

Many Americans believed it was only right that the United States, having paid for the Panama Canal, should receive more benefit from it than other countries. In their opinion, coastwise shipping should be exempt from tolls despite Great Britain's complaint that this would lead to an unfair advantage in international trade. Others felt that the toll exemption was a selfish policy that would damage the nation's reputation worldwide. US president Woodrow Wilson held the latter view, and he agreed with Britain that inequality with respect to tolls violated the treaty of 1901. He asked Congress to change the law—a politically unpopular move that his

SOURCE. Joseph P. Tumulty, *Woodrow Wilson as I Know Him*. Garden City, NY: Doubleday, 1921. pp. 162–168. Courtesy of Doubleday Broadway Publishing Group.

advisers warned would result in a split in his own party and affect its ability to pass domestic legislation. After heated congressional debate the toll exemption was repealed. In the following viewpoint Joseph P. Tumulty, a politician who served as President Wilson's private secretary, commends him for his political courage in the face of bitter criticism.

In an introduction to *The Panama Canal Tolls Controversy* [1914], edited by Hugh Gordon Miller and Joseph C. Freehoff, Mr. Oscar S. Straus wrote:

> There is no more honourable chapter in the highly creditable history of the diplomacy of our country than the repeal of the Panama Tolls Act under the present administration. Being a controversy affecting our international relations, it is gratifying that, aside from the leadership of the President, the repeal was effected not solely by the party in power, but by the help of leaders in all three parties, rising above the plane of partisan politics to the higher reaches of broad statesmanship, guided by a scrupulous regard for our international character in accord with "a decent respect for the opinions of mankind," as expressed in the Declaration of Independence.

President [Woodrow] Wilson himself, after the repealing act had been passed, remarked,

> When everything else about this Administration has been forgotten, its attitude on the Panama Tolls treaty will be remembered as a long forward step in the process of making the conduct between nations the same as that which obtains between honourable individuals dealing with each other, scrupulously respecting their contracts, no matter what the cost.

In making his recommendations to Congress he, almost with high disdain, ignored legal technicalities and diplomatic quibbles and took high moral ground. Said he,

Joseph Tumulty stands by US president Woodrow Wilson in 1919. (© **Hulton Archive/Getty Images.**)

The large thing to do is the only thing we can afford to do, a voluntary withdrawal from a position everywhere quoted and misunderstood. We ought to reverse our action without raising the question whether we were right or wrong, and so once more deserve our reputation for generosity and for the redemption of our every obligation without quibble or hesitation.

An act passed in 1912 had exempted American coastwise shipping passing through the Canal from the tolls assessed on other vessels, and the British Government had protested against this on the ground that it violated the Hay-Pauncefote treaty of 1901, which had stipulated that the Canal should be open to the vessels of all nations "on terms of entire equality." Other nations than England had an interest in this question, and there was a suspicion that some of them were even more keenly if not more heavily interested; but England took the initiative, and the struggle to save the exemption was turned, in the United States, into a demonstration by the Irish, Germans, and other anti-British elements. Innate hostility to England and coastwise shipping interests formed the backbone of the opposition to any repeal of this exemption, but the [William Howard] Taft Administration had held that the exemption did not conflict with the treaty (on the ground that the words "all nations" meant all nations except the United States), and British opposition to the fortification of the Canal, as well as the attitude of a section of the British press during the Canadian elections of 1911, had created a distrust of British motives which was heightened by the conviction of many that the Hay-Pauncefote Treaty had been a bad bargain.

> President Wilson . . . believed the exemption [of US ships from canal tolls] was in violation of the treaty.

President Wilson's Belief

It was understood early in President Wilson's Administration that he believed the exemption was in violation of the treaty, but not until October did he make formal announcement that he intended to ask Congress to repeal it. The question did not come into the foreground, however, until March 5, 1914, when the President addressed this request to Congress in ominous language, which to this day remains unexplained. "No communication I addressed to Congress," he said, "has carried with it more grave and far-reaching implications to the interests of the country." After expressing his belief that the law as it stood violated the treaty and should be repealed as a point of honour, he continued: "I ask this of you in support of the foreign policy of the Administration. I shall not know how to deal with other matters of even greater delicacy and nearer consequence if you do not grant it to me in ungrudging measure."

The first word I received that the President contemplated addressing Congress, asking for the repeal of Panama Tolls, came about in this way: I was notified after dinner one evening that the President wished to confer with me in his study. When I arrived at the White House Mrs. [Edith Bolling Galt] Wilson met me and informed me of the plan which the President had in mind with reference to this matter and of his decision to issue a statement that night which would be carried in the newspapers the following morning, and of his determination to address Congress, asking for a repeal of the Panama Tolls. Mrs. Wilson showed considerable excitement over the President's proposed step when she discussed the matter with me as I arrived at the White House. She said she had argued with the President and had tried to persuade him that if he intended to do so unusual a thing that now was the inopportune moment for it for the reason that it would create a party crisis and probably a split, the result of which we could not foresee. When I went into the President's study, he read me the announcement he had prepared for the papers. The full significance and the possible danger which lay in the proposed move that the President was about to make struck me at once. Frankly I put the whole political situation in the country before him as it would be affected by his attitude in this matter, saying to him that the stand he was about to take would irritate large blocks of Irish, Germans, and other anti-British elements in the country, and that we might expect that the leaders in our own party . . . would be found in solid opposition, and that, at a time when we needed every bit of strength to put our party programme of domestic legislation into effect, it seemed to me unwise to inject this matter, which could only be a disturbing element, into our party's councils. In discussing the matter with me, after I had presented the objections to it, which I did with great feeling and probably some irritation, he said: "I knew the view you would take of it, but, unfortunately, every argument you lay before me in opposition to the programme I have outlined in this statement is purely a partisan one and one whose value I cannot recognize at this time. I must not count the effect of a move of this kind upon my own personal political fortunes. I am the trustee of the people and I am bound to take cognizance of the fact that by reason of our attitude on Panama Tolls our treaties are discredited in every chancellery of Europe, where we are looked upon as a nation that does not live up to its plighted word. We may have made a very bad bargain with England on Panama Tolls, but it will be all the more credit to us if we stand by an agreement even when it entails a sacrifice on our part. The men who were parties to this treaty, like Joseph Choate, all agree that we have been indulging in hair-splitting and that we have done a great injustice to England. I ought not, therefore, to be afraid, because of the antagonisms that will be created, to do my duty and risk my political future if necessary in righting a great wrong. We cannot expect to hold the friendship of the world, especially of England, France, and Japan, if we are to treat agreements not as inviolable contracts, but as mere matters of convenience, whose plain terms are to be ignored when matters of

"We cannot expect to hold the friendship of the world . . . if we are to treat agreements not as inviolable contracts, but as mere matters of convenience."

expediency dictate. I know that the Irish, through the Hearst newspapers, will cry out that I have surrendered to England, that I am attempting to hand over to Europe a quasi-control over the Panama Canal. As a matter of fact, we are in bad by reason of our attitude on Panama Tolls with various leading nations of Europe, and some unforeseen contingency may arise where it will be found that the reason for their withdrawal of friendship for us was our petty attitude in this matter. I realize, as you urge, that the leaders of our party will be found in opposition, but I must forget this and try to work the matter out so that at least I shall have cleared my skirts and have done what is possible for me to do to right a great wrong."

Political Opposition

When the President concluded his statement I put before him the possible reaction against his administration and him personally which might be reflected in the returns of the Congressional elections to be held that year. He replied by saying:

> I have calculated every element in the situation and I have concluded where the path of duty lies. If we begin to consider the effect upon our own political fortunes of every step we take in these delicate matters of our foreign relations, America will be set adrift and her word questioned in every court in Europe. It is important that every agreement that America subscribes her name to shall be carried out in the spirit of those who negotiated it.

On March 5, 1914, the President addressed Congress and asked for a repeal of Panama Tolls and immediately the fierce fires of party opposition began to burn. His party leaders expressed their opposition to the repeal in open, honourable, and vigorous fashion and the fight was on. . . .

The President was adamant under the bitterest criticism. His attitude brought down on him a shower of personal abuse and vituperation. . . . Nevertheless, the tolls exemption was repealed in June. . . .

Sir Edward Grey, then [Great Britain's] Secretary of State for Foreign Affairs, commenting upon the action in the House of Commons said: "It has not been done to please us, or in the interest of good relations, but I believe from a much greater motive—the feeling that a government which is to use its influence among nations to make relations better must never, when the occasion arises, flinch or quail from interpreting treaty rights in a strictly fair spirit."

VIEWPOINT 10

The Panama Canal Project Was Conducted Less Admirably than the Popular Media Claimed

Chris Rasmussen

The following article explains how, at the time the Panama Canal was built, it represented more to Americans than an immense engineering feat that would help the world and a source of pride. They had an idealized picture of the society being established in the Canal Zone, of which the media presented only the good side. The less admirable aspects of life there, such as racism, autocratic control by the canal administrators, and a high death and injury rate among workers, were not publicized. Recently, historians have focused on these darker facts and on tensions between Americans and Panamanians, leading some to say that the canal can no longer be seen as a symbol of American ability and accomplishment.

SOURCE. Chris Rasmussen, "Water World: How the Panama Canal Shaped American Social Thinking," *Bookforum*, April–May 2009.

Chris Rasmussen teaches history at Farleigh Dickinson University in Madison, New Jersey.

In September 1913, Woodrow Wilson pushed a button in the White House, sending an electric pulse that detonated the Gamboa Dike, some two thousand miles to the southwest in the Panama Canal Zone. The dam's collapse sent water rushing from Gatun Lake into the canal, the culmination of a mammoth, decade-long construction project.

The project's payoff was still a year away, however—in 1914, the [US ship SS] *Ancon* sailed across the Isthmus of Panama, through the Miraflores locks, and into the Pacific Ocean. The new fifty-one-mile gap in the isthmus knitted the Atlantic and Pacific oceans together, a breakthrough widely hailed as the eighth wonder of the world. But it was also, and primarily, a potent symbol of the imperial ambitions of the United States. Henceforth, American power and prosperity would flow through the canal. Theodore Roosevelt reckoned its creation to be not only the most important achievement of his presidency but also the most important event in American foreign affairs since the republic's founding. The canal expanded American trade, linked the navy's Atlantic and Pacific fleets, and marked America's ascendancy as a world power.

The canal project itself involved considerable flexing of American might. Until 1903, the isthmus was part of Colombia. When negotiations between that country and the US over the proposed canal broke down, Roosevelt sent warships to back the Panamanian independence movement, then quickly framed an agreement with Panama, which had become an independent nation in November 1903. The Hay-Bunau-Varilla Treaty, ratified by the Senate in 1904, granted the United States complete control over the canal and the ten-mile-wide Panama

Canal Zone in perpetuity. As Roosevelt declared flatly in 1911, "I took the Canal Zone."

Alexander Missal's *Seaway to the Future* [2008] focuses on Americans' image of the canal's construction as a model for the orderly, conflict-free society that canal boosters hoped to create in the United States. Between 1880 and World War I, the US underwent rapid industrialization, experienced wrenching strikes, and absorbed millions of immigrants. Missal endorses historian Robert Wiebe's contention that many middle-class white Americans, fearful of a militant working class, embarked on a "search for order," seeking innovative management techniques to maintain stability in a newly industrialized, urbanized society. America's ambitious undertaking, the Panama Canal, became more than a stupendous engineering feat, also furnishing a model of a "utopian" middle-class community, in which class divisions were rendered invisible and unthreatening.

> The Panama Canal became more than a stupendous engineering feat, also furnishing a model of a 'utopian' middle-class community.

A Model Society

Missal, a journalist, focuses on the "Panama authors," who chronicled the decade-long effort to cut a channel across the country and ballyhooed the canal to the American public. These writers, he contends, were more significant than the canal itself: "For their middle-class audience, the writers' construction of the Canal was more relevant than the concrete on the Isthmus." His "alternate history" of the canal thus "takes place not on the Isthmus but within the United States—not in excavations of dirt but in people's minds." Photographers produced thousands of images of the canal, which complemented the work of the writers by offering Americans an "engineered view" of the project and a vision of a new America.

The Panama writers hailed the Canal Zone as a model society, a collectivist "mini-state" in which the government would not only oversee construction but also regulate virtually every aspect of society. The project was overseen by the Isthmian Canal Commission (ICC), which reported to the secretary of war, William Taft; day-to-day management of the zone was entrusted to a series of chief engineers: John F. Wallace (1904–1905), John F. Stevens (1906–1907), and George W. Goethals, who presided over the canal's construction from 1907 until its opening in 1914, then served for two years as the zone's first governor. (To ensure that Goethals would have complete control, [US president Theodore] Roosevelt named him chairman of the ICC, as well as chief engineer.)

Like most would-be utopias, the Canal Zone was far from democratic. Goethals, an accomplished engineer, considered the task of governing the zone's sixty thousand residents ("Zonians") both more important and more difficult than the technological feats of digging the canal and erecting its monumental locks. An autocrat, Goethals unflinchingly assumed the role of benevolent despot and strove to maintain a tightly controlled community that would furnish a model for Progressive reformers in the United States. Artist Joseph Pennell, who produced a renowned series of lithographs of the canal's construction, pronounced the zone "the best governed section of the United States." A mechanic put it differently: "There ain't any democracy down here."

Racial Discrimination

Goethals was especially determined to maintain discipline among his workforce. In order to prevent workers from developing a sense of solidarity, organizing, or striking, he (like Stevens before him) deliberately maintained sharp distinctions between white Americans, black Americans, West Indians, and Europeans. White

Americans of the Early Twentieth Century Idealized the Panama Canal Project

These excerpts from "Our Canal," a long poem by Harriet Monroe published in 1914, show the strong feelings Americans of the era had about the Panama Canal. The public—influenced by the media, which hyped the project to get large audiences just as media of today often do—considered the canal important not only for practical reasons, but also because of the idealized society to which people believed it would lead.

Here while we dig the Ditch
We would build you a perfect state,
Where service makes men great
And the great scorn to be rich;
Where each one has his place
And a measure more than his mead—
A banner of joy to grace
The strength of the daily deed;
Where wan Disease, the slayer,
Is trapped in his poison lair
With Squalor and Want and Care;
Where the Work is a marching song
Sung by us all together,
Bearing the race along
Through good and evil weather.
Oh tell them, shout it through the halls of time!—
When the Big Chief unrolls his glorious plan,
Draws hearts and hands together in perfect rhyme,
Nothing shall be impossible to Man!. . . .

We shall circle the earth around
With peace like a garland fine;
The warring world shall be bound
With a girdle of love divine. . . .

All that was writ shall be fulfilled at last.
Come—till we round the circle, end the story.
The west-bound sun leads forward to the past
The thundering cruisers and the caravels.
Tomorrow you shall hear our song of glory
Rung in the chime of India's temple bells.

O lazy laughing Panama!
O flutter of ribbon 'twixt the seas!
Pirate and king your colors wore
And stained with blood your golden keys.
Now what strange guest, on what mad quest,
Lifts up your trophy to the breeze!
O Panama, O ribbon-twist
That ties the continents together,
Now East and West shall slip your tether
And keep their ancient tryst.

SOURCE. *Harriet Monroe, "Our Canal,"* Poetry: A Magazine of Verse, *February 1914. pp. 156–158. www.poetryfoundation.org.*

and black employees were segregated. African Americans and West Indians received lower pay.

> The Panama writers created their Utopian vision of the canal by relentlessly excluding nonwhites and the specter of labor strife from their accounts.

As Missal observes, the Panama writers created their utopian vision of the canal by relentlessly excluding nonwhites and the specter of labor strife from their accounts. The writers and photographers depicted the excavation, the steam shovels, and the canal's locks in copious detail but promulgated a sanitized version of a well-ordered society by consciously discounting the aspirations of thousands of workers. Missal concludes his book with a plea that the Panama writers' fictive utopia, predicated on "exclusion and discrimination," be "shattered."

Julie Greene's *The Canal Builders* [2009] reverses the blinkered logic of these original portrayals and provides a thorough breakdown of how the canal project fell far short of its ballyhooed democratic promise. A history professor at the University of Maryland, Greene painstakingly recounts the hard, dangerous jobs of the men who dug the canal. The construction project was a "global magnet" that drew tens of thousands of workers from the United States, Canada, Europe, South America, India, and elsewhere. Half the workers were West Indians, most of them unskilled. Some five thousand workers died due to disease and construction accidents.

Stevens and Goethals deliberately sought to divide and conquer their workforce, Greene writes, by exploiting racial and ethnic divisions against any threat of growing class solidarity. Initially, skilled and highly productive workers, regardless of race or nationality, were paid in gold-backed American dollars, while unskilled laborers received silver Panamanian currency. "Gold" workers received not only higher wages but also better housing and food, annual vacations, and other perks. In

1906, Stevens ordered all nonwhite workers transferred to the silver roll; two years later, Roosevelt decreed that only American citizens (who made up between 14 and 19 percent of the workforce) could be on the gold roll. Workers were divided, and the zone's model society was riven by economic, ethnic, and racial tensions. Professionals disdained skilled laborers, and both despised the unskilled. In a notable effort to transform the zone from a rugged, nearly all-male construction camp into "a small bit of the United States," with a more contented, stable labor force, the ICC in 1906 began encouraging women to come. (There were also sensational reports that the US government had retained several hundred prostitutes for canal workers, a charge Roosevelt sought to dispel when he visited the zone in a 1906 PR offensive.)

Greene seconds Missal's view that Americans invested the canal project with great psychic significance. But while Missal describes the Panama writers' Canal Zone as intended to be a model for United States society, Greene suggests a more complex relationship between the US and the zone. The zone would offer "a display of America's domestic strengths in a world setting" and furnish a laboratory permitting reformers to experiment with activist government and expert management in ways that could improve American society and create a "progressivism for the world." Social engineering in the zone could thus aid the cause of reform both at home and abroad.

> Panamanians . . . resented the zone, which had sliced their nation in half.

Panamanian Resentment

Goethals strove to keep the zone, which he regarded a model of order, distinct from Panama—particularly from the nightlife of Panama City, which he dreaded as a source of chaos and trouble. Goethals's undisguised contempt for the country, and that of many American workers, stoked resentments that smoldered for decades. Panamanians in turn resented the zone, which had sliced their nation in half, and the Americans who had "foreignized" Panama by gaining influence over the nation's political elite.

Tensions between Americans and Panamanians erupted in 1912, as upward of one thousand American workers, soldiers, and marines boisterously celebrated the Fourth of July in Cocoa Grove, Panama City's red-light district. Fearful of violence and vandalism, barkeeps and madams shut their establishments. Undeterred, the revelers battered their way into saloons and bordellos. Ordered to quell the disturbance, Panamanian police and soldiers arrested many Americans, beat several others, and killed one. Recriminations between the Panamanian and American governments, each of which blamed the other for the riot, dragged on for three years. And as Greene observes, "For a day the Cocoa Grove riots had brought Goethals's efficient machine to a standstill."

The *Ancon*'s historic journey through the canal's locks in August 1914 was overshadowed by the war in Europe, which had broken out only days earlier. Europeans' deployment of deadly new technologies to slaughter one another by the millions shattered many Westerners' faith in progress, technology, government—and, for some, humanity itself. In 1915, the Panama Pacific International [Exposition], a world's fair held to commemorate the completion of the canal and America's newfound power, opened in San Francisco. The fair's most renowned attraction, a five-acre model of the canal, celebrated the historic achievement. As Greene observes, this model, like most published accounts of the canal's construction, depicted a world of perfect order, obscuring the privation of the workers who cleared trees, excavated soil, and blasted rock for a decade to link the Atlantic and the Pacific.

While most Americans regarded the canal as a symbol of American technological, commercial, and military prowess, it remained a source of resentment for Panamanians. In 1964, violent protests against the United States's continued control over the Canal Zone left nineteen Panamanians and three Americans dead. In the aftermath, international pressure mounted on the US to transfer control to Panama. A majority of Americans, however, opposed relinquishing the canal, and the issue vexed the nation's politics throughout the 1970s. "We bought it, we paid for it, it's ours . . . and we are going to keep it," declared Ronald Reagan in his 1976 primary campaign against Gerald Ford. Conservative Republicans' fervent determination to retain control of the canal revived Reagan's flagging candidacy, bringing him to the verge of wresting the GOP nomination from Ford, the sitting president. The Torrijos-Carter Treaties of 1977, in which the United States agreed to transfer the canal to Panama on New Year's Eve of 1999, provoked intense opposition among Americans, many of whom viewed the canal as an object of national pride and a bulwark of national security. In his victorious campaign against Jimmy Carter in 1980, Reagan castigated the "giveaway" of the canal as a vestige of Vietnam-era defeatism.

A Lost Symbol

Nearly a decade after Panamanians gained control over the canal, and nearly a century after its completion, it is now, according to Greene, a lost symbol of America's rise to global power. The canal has receded as a hallmark of American technological prowess and military might. Its construction also suggests another story, one overlooked by contemporaries, who, in much the same fashion that they overlooked its workers, failed to recognize the canal as a harbinger of a globalized market not only for goods and capital but for labor. Workers from the Americas, Europe, and Asia forsook their homes and families for

LIFE
PANAMA CITY:
AT CANAL ZONE BORDER
RIOTERS HOIST FLAG
ABOVE A FLAMING CAR
JANUARY 24 · 1964 · 25¢

jobs in the zone, inaugurating an era in which transnational labor migrations became a fixture of the global economy.

As Missal points out, the Panama Canal could inspire middle-class hopes for an America devoid of class and racial conflict only by effacing workers and nonwhites from the vision of a US-administered utopian world order. Likewise, Greene has delved into the mythology of the canal as a heroic achievement of engineers and social visionaries to highlight how the tens of thousands of workers resisted becoming cogs in Goethals's "efficient machine." These interlocking new narratives of the canal's construction not only supply a corrective to the willful blind spots of the past but also provide a useful vantage on the world bequeathed to us by the forces that set out to put America astride the globe nearly a century ago.

Photo on previous page: A 1964 cover of *Life* magazine shows rioters protesting US control of the Panama Canal. (© **Bill Eppridge/ Time & Life Pictures/ Getty Images.**)

The Panama Canal Project Was a Monument to Optimism and National Pride

Edward Tenner

The following viewpoint compares the building of the Panama Canal to the moon landing in the minds of the public of their respective eras. Both were sources of national pride, representing the height of American technological capability and military power, and both were greatly admired by other nations. Many factors enabled the United States to succeed in building the canal after the French had failed, the author says. But perhaps the most significant factor was confidence, both in the belief that the canal would be beneficial to the world and in the nation's ability to solve unforeseen problems. As was true later with the moon program, the building of the canal enjoyed a greater tolerance for risk and fewer cost-benefit analyses than would exist today. Edward Tenner is a historian of technology and culture.

SOURCE. Edward Tenner, "Digging Across Panama," *Humanities*, January–February 2011.

The Panama Canal marked a transition: from the great public works of classical Egypt, Rome, and China, built by backbreaking labor, to the far more mechanized triumphs of the twentieth century. And as a monument to optimism and national pride, only the U.S. space program of the 1960s has been its true successor. Both the canal and the moon landing were assertions of American military power and engineering prowess. Both mobilized outstanding international talent to overcome obstacles that could not be gauged fully at the outset. Each was launched by a charismatic new president in the wake of a crisis: by Theodore Roosevelt shortly after assuming office following President William McKinley's assassination, by John F. Kennedy not long after the Bay of Pigs fiasco [a failed attempt to overthrow Cuban leader Fidel Castro]. Each project subordinated economic calculation to political and military logic and was a bonanza for suppliers of state-of-the-art equipment. Each pushed the endurance of the human body to its limits, enlisting physicians alongside engineers. They aimed, respectively, at control of the sea and of space, but each also changed global consciousness: one by linking the two great oceans, the other by creating iconic images of an astronaut planting the American flag on the moon and of the earth as a fragile blue island.

Though the program does not mention the Apollo missions, the evidence of the PBS documentary *Panama Canal* suggests that for all these similarities, there were equally striking differences. In the 1960s, America was the dominant superpower, challenged in space by the Soviets and Sputnik; in the early 1900s, America was just emerging as a global force after defeating Spain. In the canal's case, though, the impetus was not the threat of a European power's success but the consequences of what was deemed the century's most spectacular technological and financial failure, the French sea-level canal project in Panama.

Chief engineer of the Panama Canal project George W. Goethals inspects spillway control points at the Miraflores Locks. (© Time & Life Pictures/ Getty Images.)

France had been America's scientific inspiration. Thomas Jefferson eagerly noted its eighteenth-century standardization of military parts, which sparked American mass production after its introduction in our own arsenals. Our first academic engineering program was created at the U.S. Military Academy at West Point by its superintendent, Sylvanus Thayer, a civil engineer who revered French technical education and reformed the curriculum on the model of his other alma mater, the Ecole Polytechnique (still under the French Ministry of Defense today) with compulsory French instruction. Ferdinand de Lesseps, who had become a world hero of civil engineering as well as a national icon as the builder of the Suez Canal, believed he could repeat his success with a sea-level canal across the Isthmus of Panama. He discovered too late the exceptional difficulties of the site:

unstable soil, overpowering vegetation, virulent diseases, and especially the turbulent Chagres River. Over eight years, twenty thousand lives were lost to work hazards, malaria, and yellow fever. There was little to show for an investment of more than a billion francs, the equivalent of $280 million today. By the time the project collapsed, the eight hundred thousand shareholders had lost their savings, the greatest crash of the nineteenth century, according to Matthew Parker's *Panama Fever* [2008].

> The Hay-Bunau-Varilla treaty . . . reflected a widespread western belief in benevolent imperialism.

While de Lesseps never recovered professionally or psychologically, his chief engineer, Philippe Bunau-Varilla, who had a personal fortune at stake, refused to abandon the project, even after a new Panama Canal company was ready to give up and American politicians and strategic planners were leaning toward a canal through Nicaragua instead. He allied himself with the original superlobbyist and master fixer, the attorney William Nelson Cromwell. Then, as sole representative of the newly independent nation of Panama (freed from Colombia by a revolution protected by American warships), Bunau-Varilla signed a treaty that gave the United States virtual sovereignty in the Zone and a free hand to build the canal.

American Self-Confidence

The Hay-Bunau-Varilla treaty was one-sided, even by the standards of its time. But it reflected a widespread western belief in benevolent imperialism. Theodore Roosevelt wrote to a friend at the British Foreign Office that just as British rule in Egypt and India benefited those countries and the world, so American hegemony would be good for the Panamanians. Woodrow Wilson later declared that his own military interventions would "teach the South American Republics to elect good men."

A British Writer's 1913 Speculation on the Impact of the Canal

Some readers may perhaps think that these forecasts of the results of running a canal through the isthmus of Panama are somewhat exaggerated. It is sufficient to point out to such a critic how different the course of American and world history might have been if Nature had left a practicable channel between the two Americas. The effect of erecting an artificial passage there in these days may be even greater than at present we can imagine. Some of these results will be apparent at once; others may take decades or even centuries to materialize. . . . If a dozen experts were, in 1914, to write out and place in the libraries of the British Museum and of [the US] Congress their respective forecasts bearing on this subject, sealed up and not to be opened till A.D. 2000, they might make curious reading in that year. . . .

As the past four hundred years have belonged to the Atlantic, the present century and others to come may belong to the Pacific. That area of 70,000,000 square miles may become the main theatre of the rivalries commercial, political, and racial of the most powerful nations of East and West. Some believe that the world is advancing to that loud and fateful day when East and West will fight out their long difference in some naval and aerial Armageddon on and above this miscalled Pacific. Without straining our imaginations to this extent, we may well observe that the canal brings Eastern and Western civilizations into much closer contact and competition than before. . . .

I do not wish to end this little book with presages of future discord. We must all hope that the Panama Canal will prove a new and powerful influence for peace, that it will bring even East and West together, not in strife and suspicion, but in friendship and a better mutual understanding. There is surely a human interest and sympathy transcending even those racial divisions which may seem most insuperable. The great nation which has given this splendid gift to the world should ask no better or more selfish reward than that it may contribute to the welfare and progress of humanity at large.

SOURCE. *J. Saxon Mills,* The Panama Canal: A History and Description of the Enterprise. *London: Thomas Nelson and Sons, 1913, pp. 316–317, 322. www.openlibrary.com.*

Among early twentieth-century great powers, and some smaller ones, despite voices of dissent, arrogance was ecumenical.

The open projection of power would not be enough to build a canal where the French had failed. Another attitude was needed: promethean self-confidence. When the treaty was signed, nobody knew how to solve problems that had defeated the French. The majority of the Board of Consulting Engineers, an expert international committee, recommended a sea-level canal like the one de Lesseps had attempted. While the force of the Chagres River would have overwhelmed it and doomed the project, proponents of lock construction realized their own solution also had disastrous precedents, including the failure of the Johnstown, Pennsylvania, dam and a vessel's collision with a lock on Britain's Manchester Ship Canal.

Perhaps only a state enterprise backed by military methods could have solved such open-ended problems of building the canal. America's experience in building transcontinental railroads across inhospitable terrain certainly helped. The first head of canal construction, John Wallace, was a railroad engineer attracted by a salary second only to the president's on the federal payroll. It was John Stevens, an even more notable railroad builder, whose observations of the realities of the site's greatest challenge, the Culebra Cut, led him to convince Roosevelt to back the lock idea, a minority viewpoint. But the ultimate direction of the project fell to a career engineering officer, George Goethals. Construction was a paramilitary operation: Although paternalistic, the venture was enlightened by the standards of the day.

Factors in the Project's Success

How did the United States succeed where perhaps the most admired engineer of the nineteenth century had failed? Timing was part of the answer. France was a world leader in medicine, but it was Walter Reed, a U.S.

Army physician, who recently had eradicated yellow fever in occupied Havana by preventing mosquito breeding, applying decades of studies by the brilliant Cuban physician and epidemiologist Dr. Carlos Finlay. Another army doctor, William Gorgas, relentlessly used the same method to suppress mosquitoes in the Canal Zone, eliminating a major source of the French project's high mortality rate. Bucyrus steam shovels, recently developed, moved earth three times as fast as French equipment. Advances in electrical engineering by the 1890s made possible hydroelectric-powered, motorized lock systems with controls that performed superbly for decades. The mass-circulation newspapers and magazines that began to flourish in the mid-1890s stoked popular enthusiasm.

The project benefited from more troubling changes too, especially from the drop in world sugar prices that put pressure on wages in Barbados, a preferred source of labor. The canal was staffed according to blatant racial and ethnic stereotypes. Chinese laborers, initially favored for their heroic work in railroad building, were considered too ambitious. Once they arrived, they might leave digging for shop keeping. American laborers, white and black, as well as Europeans, were thought too likely to agitate and strike. West Indians were favored for the most arduous and dangerous jobs; the ten cents an hour the canal offered surpassed the dwindling wages of Caribbean plantations. White managers and skilled workers were paid in gold, the West Indian laborers in silver specie [money] that was also a euphemistic metaphor for the color line.

For many viewers of the *American Experience* [television] program, one of the great surprises may be the extent of high-handed management by Goethals. Unlike some later "czars," he actually could rule autocratically, dominating Canal Zone life as well as the actual construction work. He set his tone early by refusing to negotiate with some of the most skilled and apparently indispensable workers, the steam shovel operators, effectively shutting down the project until strikebreakers arrived to replace them. His ruthless drive was most apparent in the nonstop excavation of the canal's most daunting section, the Culebra Cut, where landslides constantly imperiled all workers on the site, but especially the West Indians, whose living conditions remained miserable. But the men of the "Gold Roll" who did not challenge his authority or the wage scale enjoyed increasingly comfortable accommodations, reversing the initially high turnover and persuading families to join their men. For them, the military strongman created a tropical welfare state that persisted well into the new century.

Overcoming the Unexpected

The Panama Canal illustrates the principle that the economist Albert O. Hirschman has called the Hiding Hand. People begin many enterprises because they don't realize how difficult they actually are, yet respond with ingenuity that lets them overcome the unexpected, as the Apollo program's engineers and astronauts were later to do. The testimony in *Panama Canal* also shows the power of the heroic image of technology in the early twentieth century. It was felt even by the exploited laborers, who still shared the nineteenth century's stoic approach to industrial risk. Three percent of white American workers and nearly 14 percent of West Indians died. Despite improvements in sanitation, it was "a harsh nightmare," the grandson of one of those workers declares in the program, but he also recalls the pride of his grandfather in participating in one of the world's great wonders. In fact, many returnees were inspired by their achievement to join movements for greater economic and political equality

> Many [workers] were inspired by their achievement to join movements for greater economic and political equality in the 1920s and 1930s.

in the 1920s and 1930s, the roots of the decolonization movement.

No change to the world's landscape since the Panama Canal—the Interstate Highway System, the Channel Tunnel, or China's Three Gorges Dam—has come dose to its acclaim. It has kept its magic even as the other great technological breakthrough of its time, aviation, has passed from inspiration to regimented mass routine.

It was conceived without today's cost-benefit analysis. The historian Walter LaFeber acknowledged in 1978, during often emotional controversies over new U.S.-Panama treaties, that "[w]hether the investment [of $400 million in the Canal] has been amortized is an unanswerable question . . . like a thousand-piece jigsaw puzzle which has important pieces missing." Members of Congress, he continued, had really wanted to subsidize shippers and never thought about recovering construction costs.

A sea-level alternative through Nicaragua might well have made more economic sense, but overcoming the landslides, the flooding, the heat, and the mosquitoes that threatened to doom completion, as documented by print mass media at their height, made construction in Panama uniquely audacious. Perhaps our aspirations both on earth and in space are more modest now because we are more rational in our ability to foresee costs, especially environmental ones.

Planners of any future grand ventures should bear in mind the words of Joseph Gavin, in charge of the original lunar module of the space program, applicable to all the canal's builders from de Lesseps to Goethals: "If a project is truly innovative, you cannot possibly know its exact cost and exact schedule at the beginning. And if you do know the exact cost and the exact schedule, chances are that the technology is obsolete." Or as global student graffiti was proclaiming in 1969, year of Apollo 11: "Be realistic. Demand the impossible."

CHAPTER 3

Personal Narratives

First Lady Helen Taft Recalls Her Trips to the Panama Canal Zone

Helen Herron Taft

Helen Herron Taft was the wife of US president William Howard Taft, who was the US Secretary of War (in charge of military affairs, even in peacetime) at the time of their first visit to Panama. In the following excerpt from her autobiography, Mrs. Taft tells about the trip they took when construction of the canal was just beginning. She describes the warm welcome they received from the Panamanians, both the officials they visited and the crowds that cheered them when Taft spoke in public. Everyone was enthusiastic about the canal, and Mrs. Taft was impressed by the natural beauty of Panama, which she says she will long remember.

Photo on previous page. Men and women pose at an inspection party during construction of a Panama Canal lock. (© **Bettmann/ Corbis/AP Images.**)

SOURCE. Helen Herron Taft, *Recollections of Full Years*. New York: Dodd, Mead and Company, Inc., 1914, pp. 283–292. Courtesy of Dodd, Mead and Company, Inc.

I had hardly got my house on K Street [in Washington, DC] in order before something happened in Panama which made it imperative for the Secretary of War to go down to the Isthmus and give the situation his personal attention. There was a state of popular discontent among the Panamanians complicated by question of zone boundaries, jurisdiction, postal regulations, tariff inequalities and a few other matters, and by that time we had too much at stake in the Canal Zone to risk long distance or dilatory regulation. . . .

I was very glad to have an opportunity to see the beginning of what I knew was to be the greatest enterprise ever undertaken by the United States, so I fully approved of my husband's suggestion that I accompany him to Panama. . . .

On this trip Mr. [William Howard] Taft went to Panama as a representative of the President of the United States [Theodore Roosevelt] for the purpose of presenting to the President of Panama [Manuel Amador] a message of friendship, and to make, if possible, an amicable adjustment of the differences between American and Panamanian interests.

It was in November, 1904. . . .

We sailed on the little *Dolphin* from New Orleans to Pensacola, where the cruiser *Columbia* lay waiting to take us down to Panama, and it was to the boom of saluting guns, the cheers of hospitable Pensacola citizens and the strains of "The Star Spangled Banner" that we got under way on this first memorable trip to the Canal Zone.

We arrived at [the Panama port city of] Colon on a Sunday morning, and I remember distinctly that it seemed more like "getting home" than like getting to a strange place. The whole atmosphere and surroundings, the people, the language they spoke, the houses and streets, the rank earth odours and the very feel of the air, reminded me so strongly of the Philippines [where Taft had been Governor] as to give me immediately a delightful sense of friendly familiarity with everything and everybody.

We were met at Colon by the vice-President of Panama, Señor [Ruben] Arosemana, and a number of other Panamanian officials, by General [George] Davis, then Governor of the Canal Zone, and by Mr. John Barrett, the American Minister to Panama. A private train was waiting to take us across the Isthmus and we lost no time in getting started. Our visit had been "programmed" almost to the last hour of our time, and the first event was to be an exchange of formalities between the Secretary of War and the President of Panama that very afternoon. . . .

Panamanian Hospitality

Our Minister, Mr. Barrett, had a charming house in the old tropic city and on the Monday evening after our arrival he gave a dinner at which were gathered many high officials of the Panama Republic as well as all the interesting Americans who were then directing our great Canal building enterprise. . . . Many Army officers and Cabinet Ministers in full regalia and many decorations, with their wives, were seated in order of rank along the sides of the great table, which, laden with flowers and gleaming glass and candles, made a picture long to be remembered even by one whose memory is overcrowded with dinner-party scenes.

"When we arrived in Panama we were not at all certain that we should find the country in a state of tranquility."

The formality of this occasion, however, began and ended with its costuming and its beautiful tropic "setting." Nearly everybody, including the President and Madame Amador, spoke English as well as Spanish, and the evening was gay from the outset. . . .

When we arrived in Panama we were not at all certain that we should find the country in a state of tranquility; nor did we exactly; though by prompt action the

President had nipped a budding revolution only a short time before. Hostilities had been averted, but the people were in a bad temper and it was thought best to keep them "merely guessing" while the negotiations between the Secretary of War and the Panamanian government were in progress, and much of Mr. Taft's time, therefore, was spent behind the closed doors of President Amador's council chamber.

In the meantime I made myself familiar with the wonderful American project which brought the Panama Republic into existence. The Canal then was a sorry sight. The public clamour in the United States was for "making the dirt fly," but it did not look to a mere layman as if we could ever make it fly fast enough or in sufficient quantities to really bring the two oceans together. All along the line of operations the old French machinery lay buried in pathetic ruin in a tropic jungle which had all but effaced the evidences of the French enterprise, and such conditions of general unhealthiness prevailed as made it seem almost too much to expect that any kind of clean-up programme could be made effectual.

> The Panamanians are nothing if not expansively hospitable.

But all that story has been told; told in actual accomplishment with which all the world is familiar. I am only glad that I saw Colonel Gorgas and his men in that initial and contagious enthusiasm which, being sustained, resulted in a record of which we are all so proud.

The Panamanians are nothing if not expansively hospitable. On the 4th of December, after we had been on the Isthmus a couple of weeks and while the results of the official negotiations were still, as far as any one knew, "in the lap of the gods," an ocean steamship was chartered by a company of hosts, and about three hundred guests, the élite of the whole republic, were invited for a picnic party to the Pearl Islands in the Bay of Panama, and a

US first lady Helen Taft traveled with her husband, William Howard Taft, to Panama in 1904 when he was the secretary of war. (© **AP Images.**)

sail out into the Pacific Ocean. It was an all-day expedition and included the exploration of the beautiful little group, some pearl-diving for our especial benefit, a most amazing luncheon, and a dance on deck, to the music of a stringed band in gay and most decorative uniforms, at which Mr. Taft made a tremendous "hit.". . .

Demonstrations of Enthusiasm

When we returned from the Pearl Islands to Panama City that evening we were met by a pleasing surprise. The text of the agreement which had been reached by Mr. Taft and the government of Panama, and which had not been mentioned by anybody all day, had been made public during our absence and newsboys were crying "extras" in all the streets, while excited groups stood about here and there wreathed in smiles and talking with great animation. Everybody seemed wholly satisfied and wherever we went we were met with cheers and cries of "Viva!"

The keynote sounded in this agreement was that justice should be done at Panama. Mr. Taft interpreted the treaty between our country and the infant republic; he adjusted the differences with regard to postal regulations and the tariff; he defined the harbour boundaries; and, much to the satisfaction of the Panamanians, he kept within the hands of the American authorities all matters pertaining to the public health. The people realised the necessity for a pure water supply, for sewer systems, clean and well-paved streets, and the elimination of the dread diseases which made the Isthmus a death trap for white men, and all these things America offered to Panama as a free gift that the great work of building the Canal might go on. If vociferous cheers are an expression of gratitude the people were grateful.

The next day a great demonstration took place in Cathedral Plaza. We stood on a balcony of the Grand Central Hotel, on one side of the square and opposite the Cathedral, and looked out across a veritable sea of moving, swaying, white-clad humanity. As far as one could see in every direction there were people, and when Mr. Taft stepped to the balcony rail to address them they gave voice to a cheer which made it seem certain that all cause for quarrel between us had happily been removed.

There had been vague rumours that the deposed War Minister of the Panama government who had attempted to start the rebellion would, with his followers, take this occasion to make a hostile demonstration, but he was evidently sensible enough to realise that his was an unpopular cause. Moreover, his original army of two hundred and fifty men had been reduced to twenty-five, and if that were not discouraging enough he had only to contemplate the natty American Marine corps in the Canal Zone and the Pacific squadron, including the *New York,* the *Boston,* the *Bennington* and the *Marblehead,* lying out in Panama harbour unobtrusively but very positively guaranteeing peace.

> Mr. Taft . . . was imbued with the spirit of conquest as represented in our Panama Canal enterprise.

Mr. Taft in his speech to the Panamanians was earnest. He was imbued with the spirit of conquest as represented in our Panama Canal enterprise. It was to be a conquest of nature's own forces in their most formidable aspects and he expressed a determination to enforce, during his term of administration, the laws necessary to make that conquest possible, and capping all his promises of fair treatment to the people of Panama, he emphasised an insistence upon orderly government in the little republic which brought forth round after round of applause. . . .

In deciding the questions which came to him, it was necessary for him to visit the Canal seven times in as many years, and I went with him on three of his visits. The contrast between the Canal when we first visited it and were the guests of Mr. [John] Wallace, the first engineer, and as it was when we were the guests of Colonel [George] Goethals [the chief engineer] in 1912, when the Gatun Lake was more than half filled and nothing but the slides in the Culebra Cut remained for excavation, it is most interesting to look back upon. I was twice the guest of Colonel and Mrs. Goethals, and the beautiful view of the Canal Valley from the windows of their house in the town of Culebra, which has now disappeared, will long remain in my mind.

President Theodore Roosevelt Describes His Impressions of the Canal's Construction

Theodore Roosevelt

The following is a letter from US president Theodore Roosevelt to his teenage son Kermit written aboard ship during his return from a trip to Panama in 1906. He tells how he thought about the history of the discovery and settlement of the isthmus by European explorers and about the failure of the French attempt to built a canal there. He says how impressed he was by the efficiency and energy of the Americans on the job. He also expresses his enthusiasm both about the work being done and about the beautiful forest and wildlife that he observed. Theodore Roosevelt was the President of the United States from 1901 to 1909.

SOURCE. Theodore Roosevelt, "What the President Saw at Panama," in *Theodore Roosevelt's Letters to His Children*, ed. Joseph Bucklin Bishop. New York: Scribner's, 1919, pp. 179–184. Courtesy of Scribner's.

Our visit to Panama was most successful as well as most interesting. We were there three days and we worked from morning till night. The second day I was up at a quarter to six and got to bed at a quarter of twelve, and I do not believe that in the intervening time, save when I was dressing, there were ten consecutive minutes when I was not busily at work in some shape or form. For two days there [were] uninterrupted tropic rains without a glimpse of the sun, and the Chagres River rose in a flood, higher than any for fifteen years; so that we saw the climate at its worst. It was just what I desired to do.

It certainly adds to one's pleasure to have read history and to appreciate the picturesque. When on Wednesday we approached the coast, and the jungle-covered mountains looked clearer and clearer until we could see the surf beating on the shores, while there was hardly a sign of human habitation, I kept thinking of the four centuries of wild and bloody romance, mixed with abject squalor and suffering, which had made up the history of the Isthmus until three years ago. I could see [Spanish explorer Vasco Núñez de] Balboa crossing at Darien, and the wars between the Spaniards and the Indians, and the settlement and the building up of the quaint walled Spanish towns; and the trade, across the seas by galleon, and over land by pack-train and river canoe, in gold and silver, in precious stones; and then the advent of the buccaneers, and of the English seamen, of [Sir Francis] Drake and [Martin] Frobisher and [Sir Henry] Morgan, and many, many others, and the wild destruction they wrought. Then I thought of the rebellion against the Spanish dominion, and the uninterrupted and bloody wars that followed, the last occurring when I became President; wars, the victorious heroes of which have their pictures frescoed on the quaint rooms of the palace at Panama city, and in similar palaces in all capitals of these strange, turbulent little half-caste civilizations. Meanwhile the Panama railroad had been built by Americans over a half century ago, with appalling loss of life, so that it is said, of course with exaggeration, that every sleeper laid represented the death of a man. Then the French canal company started work, and for two or three years did a good deal, until it became evident that the task far exceeded its powers; and then to miscalculation and inefficiency was added the hideous greed of adventurers, trying each to save something from the general wreck, and the company closed with infamy and scandal.

Admirable Workers

Now we have taken hold of the job. We have difficulties with our own people, of course. I haven't a doubt that it will take a little longer and cost a little more than men now appreciate, but I believe that the work is being done with a very high degree both of efficiency and honesty; and I am immensely struck by the character of American employees who are engaged, not merely in superintending the work, but in doing all the jobs that need skill and intelligence. The steam shovels, the dirt trains, the machine shops, and the like, are all filled with American engineers, conductors, machinists, boiler-makers, carpenters.

> Men are changing the face of the continent . . . and the effect of their work will be felt while our civilization lasts.

From the top to the bottom these men are so hardy, so efficient, so energetic, that it is a real pleasure to look at them. Stevens, the head engineer, is a big fellow, a man of daring and good sense, and burly power. All of these men are quite as formidable, and would, if it were necessary, do quite as much in battle as the crews of Drake and Morgan; but as it is, they are doing a work of infinitely more lasting consequence. Nothing whatever remains to show what Drake and Morgan did. They produced no real effect down here, but Stevens and his men are changing the face of the continent, are doing the greatest engineering feat of the ages, and the effect of their work will be felt while our civilization lasts. I went over everything that I could possibly go over in the time at my disposal. I examined the quarters of married and single men, white men and negroes. I went over the ground of the Gatun and La Boca dams; went through Panama and Colon, and spent a day in the Culebra cut, where the great work is being done. There the huge steam-shovels are hard at it; scooping huge masses of rock and gravel and dirt previously loosened by the drillers and dynamite blasters, loading it on trains which take it away to some dump, either in the jungle or where the dams are to be built. They are eating steadily into the mountain, cutting it down and down. Little tracks are laid on the side-hills, rocks blasted out, and the great ninety-five ton steam-shovels work up like mountain howitzers until they come to where they can with advantage begin their work of eating into and destroying the mountainside. With intense energy men and machines do their task, the white men supervising matters and handling the machines, while the tens of thousands of black men do the rough manual labor where it is not worth while to have machines do it. It is an epic feat, and one of immense significance.

A Tropic Forest

The deluge of rain meant that many of the villages were knee-deep in water, while the flooded rivers tore through the tropic forests. It is a real tropic forest, palms and bananas, breadfruit trees, bamboos, lofty ceibas, and gorgeous butterflies and brilliant colored birds fluttering among the orchids. There are beautiful flowers, too.

All my old enthusiasm for natural history seemed to revive, and I would have given a good deal to have stayed and tried to collect specimens. It would be a good hunting country too; deer, and now and then jaguars and tapir, and great birds that they call wild turkeys; there are alligators in the rivers. One of the trained nurses from a hospital went to bathe in a pool last August and an alligator grabbed him by the legs and was making off with him, but was fortunately scared away, leaving the man badly injured.

> It was curious to think that in a few years great ships would be floating in water 100 feet above where we were.

I tramped everywhere through the mud. Mother did not do the roughest work, and had time to see more of the really picturesque and beautiful side of the life, and really enjoyed herself.

P.S. The Gatun dam will make a lake miles long, and the railroad now goes on what will be the bottom of this lake, and it was curious to think that in a few years great ships would be floating in water 100 feet above where we were.

A Popular Author Reports on the Dangerous Job of Cutting Through a Mountain

Hugh C. Weir

In the following narrative from his book on the Panama Canal, Hugh C. Weir describes being taken on a sightseeing trip to see the Culebra Cut, where hundreds of men were cutting a passage through the mountain. Traveling in a motor car that ran on the railroad track, he saw them at work with dynamite and steam shovels. He describes the dangers the workers faced and writes about the immense power of one particularly large steam shovel that moved a boulder while he watched. Weir was a popular author of many books, some for teens and others for the general public.

SOURCE. Hugh C. Weir, *The Conquest of the Isthmus: The Men Who Are Building the Panama Canal—Their Daily Lives, Pain, and Adventures*. New York: Penguin Putnam, Inc., 1909, pp. 83–92.

The cheery beams of the tropical sun were already streaming into my room at the Tivoli when the jangle of the alarm-clock aroused me. The sun and the Americans are early risers in Panama, I was to discover. It is during the morning that by far the greater share of the work is done on the Isthmus.

From Panama to Culebra is a distance of nearly twelve miles. The Panama Railroad does not endeavor to cope with the speed of the trains at home, and it was three quarters of an hour after our start before I alighted at the Culebra station.

As I glanced uncertainly around me, a young man with a cheery voice hailed me from the edge of the platform.

"I am Major [David du Bose] Gaillard's secretary," he announced. "If you will come with me, I'll take you over to the car." Major Gaillard is the division engineer, in charge of the Culebra section of the Canal.

> Over one hundred locomotives were backing and switching, often barely grazing each other as they darted to and fro.

It was in a cubby-hole of a garage, bordering the line of rails, that we found the motor. Peterson, chauffeur, machinist, and all-around man of information, was awaiting us with a broad grin of welcome.

"Ready?" I asked.

"In two minutes," he replied. "Here, Jim!" he called sharply, and his black boy sprang to his side.

"Turn-table!" continued Peterson briefly. In a moment a series of cleverly grooved planks were joined together in front of the motor, and the car was run onto a neatly revolving wooden track. As smoothly as clock-work, it swung around and the motor was gently deposited on the railroad, ready to spring forward at the touch of a lever.

"All in?" called Peterson curtly over his shoulder. Bending forward, he switched the starting gear into place, the car darted swiftly down the rails, and we were off.

A Motor Car on the Rails

We swung past the Culebra station, every moment adding to our speed. The morning air was whipping our faces with a pleasant, invigorating thrill, and the spirit of the swift dash was beginning to take firm hold of us.

Peterson slackened his pace as we neared the switch beyond and hurriedly crossed his hands out over the side of the car—the silent signal to the workmen ahead that he wished to change into the adjoining line. On his side Jim was duplicating the action, which was soon to grow familiar to us.

It was at Whitehouse, a small dot of a station between Culebra and the neighboring point of Empire, that we again swerved our course, and Peterson cried, "We are entering the big Cut now. Keep your eyes open, Jim!"

The need of the warning was soon apparent. Within the space of the next eight miles over one hundred locomotives were backing and switching, often barely grazing each other as they darted to and fro in the swirling mist of their own steam. A collision might come at any moment even with the experienced hand of Peterson guiding us.

Attached to scores of the engines were long rows of dirt cars, partially filled, every moment adding to their great loads of clay and rocks.

"There are more men killed on the Panama Railroad dodging dirt trains than from any other cause," Peterson grimly informed us. Which was pleasant intelligence, as we darted down past the bumping rows of swaying cars and they darted down past us, the motor and the trains really playing an exciting game of "hide-and-seek" or "prisoner's base."

"The towering walls of Culebra Cut were now rising above us, their great, rugged faces seeming to scowl in baffled rage"

Culebra Cut

The towering walls of Culebra Cut were now rising above us, their great, rugged faces seeming to scowl in baffled rage at the army of sweating men below, who day by day were ploughing deeper and farther into their sides.

For centuries these great, swollen mountains had defied the assaults of men, laughing at the efforts to bore a passage through their rocky ridges. And now the men, in their turn, were laughing at the efforts of the bullying mountains to check their advance.

Have you ever studied the picture of a noted battle-field? Do you recall the thick clouds of smoke, the spurting cannon, the stacks of rifles, the heaps of dead and wounded men?

Change the field of battle to Culebra Cut. You will see the same thick, black clouds of smoke. Instead of the belching cannon, you will find a hundred times more deadly instrument in the giant dynamite blasts. The monster steam shovels, the great levellers and air-drillers are the weapons of warfare, and the opposing forces are the armies of man and Nature. It is not one battle, but a series of daily battles, and they are all to the death.

> The monster steam shovels, the great levellers and air-drillers are the weapons of warfare, and the opposing forces are the armies of man and Nature.

Hundreds of men, thousands of men are before and behind and around us—black men, white men, yellow men, red men—men with their coats and shirts and collars off, with grimy hands and perspiring faces and straining shoulders—men to whom a dozen different languages might be addressed without finding their native tongues.

Over all tower the great, scowling cliffs, before you is the constant swirl of brown smoke, and on every side the screech of shrill locomotive whistles, the hoarse shouts of toiling men, the grinding crunch of the steam shovels.

A Monster Steam Shovel

Peterson turned suddenly as we worked our way in between the overhanging cliff walls—our speed was now little more than a bare, zig-zagging crawl—and cried crisply, "We're coming to one of the largest steam shovels on the Isthmus. Do you want to stop?"

In answer to my nod, the motor paused and I sprang out onto the ground, at close quarters with my first steam shovel.

If you can imagine pounds magnified to tons and can conceive of a monster iron scoop that can handle these tons as easily as you can handle an ordinary baseball, if you can picture such a gigantic machine so cleverly constructed that it is possible for one man to swing the great dipper where and when he pleases, you will have a dim framework of the American steam shovel as it is operating at the "Big Ditch." Can you go a step farther and imagine the man placed in such a position that he is hidden from view, the monster scoop seeming to work of its own accord—a great, rough creature of iron and steel suddenly given the power of life?

If you can, you will have an even better idea of what the steam shovel really is.

I clamored across onto the half-filled dirt train beside the motor that I might get a closer view. As I did so, the iron dipper struck a mammoth boulder half buried in the red clay below.

Deeper and deeper, its four iron teeth worked their way into the sticky mud at the base of the great stone. The boulder suddenly leaned over under the weight of the scoop and then, as I gasped, it was lifted bodily from the ground—wedged tightly between the gaping iron jaws. The shovel gave a terrific upward jerk, and almost before I realized it, the huge stone was being suspended in the air above my head.

"Jump!" shouted Peterson from the motor. "Great Scot, man, that boulder weighs twenty tons!"

I didn't wait for additional explanation. With the most rapid side step I think I have ever made, I sprang to the ground, and none too soon.

Steam shovels and trains assist workers with the dangerous job of cutting through a mountain at the Culebra Cut. (© **Hulton Archive/Getty Images.**)

The next moment the flap of the dipper opened, and the boulder dropped into the flat-car with a dull thud.

But it didn't stay. Hardly had it settled on the clay when it turned on its side, rolling ponderously toward the ground.

The steam shovel wasn't idle, however. With a slow awkward movement it again swung around, its iron edge striking the rock with a force that carromed it sharply over in the other direction. And then, as though the boulder were suddenly fired with electric energy, it plunged off toward the opposite end of the car, every instant gathering new force. Again the steam shovel worked around, and this time with a resounding jar dealt the giant slab of granite another blow.

The boulder's course was abruptly checked, but only momentarily. A third time it commenced to roll, plunging toward the ground with even greater velocity than before.

It was a thrilling crisis for the layman.

With a jerk as though it had gathered all of its energies for a final spurt, the great shovel pivoted about, hesitated as if measuring the most effective spot at which to strike, and literally grappled with its granite opponent.

The boulder's massive strength drew a ringing crash from the iron dipper, but it had brought up against an obstacle it could not move. It was vanquished. Slowly the steam shovel withdrew, hovering in the neighborhood a moment, ready for another attack. But the stone was firmly lodged this time. The dipper had done its work well.

Miles of Track

"How many miles of track would you guess have been laid on the Isthmus?" queried Peterson as our car threaded its way beyond a more than usually active row of dirt trains.

"Possibly a hundred," I suggested.

He laughed as he shook his head.

"You will have to multiply that number by four—and then add some," he rejoined. "There are 448 miles of rails in Panama—in a distance of just forty-seven miles. In other words, we often have twelve and fifteen tracks in a row. There are fully this many before us now."

During a lull in the activity around me, I glanced at the cliff above. Its scarred, jagged surface showed nearly every color of the rainbow. Here was a surface of gray, there a bright scarlet hue, yonder a line of tan, to the left a dark slate color, below a flaring yellow—the blending outlines of the different strata of dirt uncovered in the ever-deepening excavations.

> Assuredly [Spanish explorer Vasco Núñez de Balboa] would have termed the steam shovel a fabled giant lurking in the Panama wilderness like the dragons of old.

"There is the famous Gold Hill to our left," explained Peterson.

"Gold Hill?" I repeated.

"The point from which [Spanish explorer Vasco Núñez de] Balboa discovered the Pacific," the chauffeur added. "The ocean is twelve miles from here," he continued. "You can see it easily on a clear day—but I wouldn't care to have had Balboa's trip to reach it, eh?"

I wondered curiously what the explorer's feelings would have been could he have pictured the present scene in Culebra Cut. Assuredly he would have termed the steam shovel a fabled giant lurking in the Panama wilderness like the dragons of old.

Speeding Through the Jungle

It was easy to see now that we were approaching the end of the great Cut. The cliffs had broken off sharply, and the number of workmen had abruptly lessened. Peterson brought the car to a sudden halt, the turn-table was again brought into play, and we were switched off at right angles to begin our circling way back to Culebra through the dark undergrowth of the jungle.

Faster and faster grows our speed. Now the wheels seem fairly to be burning into the rails, and we tear through the heart of the wilderness at a speed that is eating up forty-five miles an hour! There is a clear track ahead, our white flags are held stiff against the breeze, and for the moment the thrill of the wild dash is our only sensation.

At our side the trees shoot by like a solid wall—we catch a confused glimpse of an occasional monkey chattering impudently at us, the sweep of a bright-plumaged bird, and the flash of a flower, but we have no time for details. We almost feel that we are devouring space.

It is barely half-past ten, and we are scheduled to board the 10.45 train for the Gatun Dam, thirty miles down the line. This thought is in Peterson's mind, as he jerks out:

"We'll make it nicely!" And we did.

A Canal Zone Policeman Reflects on the Distress of Panamanians Displaced by the New Lake

Harry A. Franck

In 1912 travel writer Harry A. Franck worked for several months as a Panama Canal Zone policeman. In the following narrative he writes about a boat trip he took on Gatun Lake when it was filling, partly covering the trees so that only their tops were above the water. The growing lake also covered native villages. Their inhabitants had been paid for the land, but they did not really believe it would be inundated and, in some cases, refused to leave until their homes were burned down to force them out. Franck says that it was incomprehensible and upsetting to them when strangers demanded that they leave their ancestral land. Franck was the

---

SOURCE. Harry A. Franck, *Zone Policeman 88: A Close Range Study of the Panama Canal and Its Workers*. Watkins Glen, NY: Century House, 1913, pp. 295–308. Courtesy of Century House Publshing.

author of more than twenty travel books describing his adventures around the world.

Dr. O—— of the Sanitary Department had gone up into the interior along the Trinidad river to hunt mosquitoes. Why he went so far away for them in this season was hard to understand. There he was, however, and the order had come to bring him back to civilization. The execution thereof fell, of course, to my friend B——, who to the world at large is merely Policeman No. ——, to the force "Admiral of the Inland Fleet," and in the general scheme of things is a luckier man than Vanderbilt [a famous, extremely wealthy family] to have for his task in life the patrolling of Gatun Lake. B—— invited me to go along. . . .

I was at the appointed rendezvous promptly at seven, a pocket filled with commissary cigars. Strict truthfulness demands the admission that it was really eight, however, when B—— came wandering down the muddy steps behind the railroad station, followed by a black prisoner with a ten-gallon can of gasoline on his head. When that had been poured into the tank, we were off across the ever-rising waters of Gatun Lake. For Gatun police launch is one of those peculiar motor-boats that starts the same day you had planned to.

Through a Forest in a Motor Boat

It was such a day as could not have been bettered had it been made to order, with a week to think out the details—a dry-season day even to the Atlantic breeze that goes with it, a sort of Indian summer of the rainy season; though the heavy battalions of gray clouds that hung all around the horizon as if awaiting the order to charge warned the Zone to make merry while it might, for tomorrow it would surely rain in deluges. The lake, much higher now than in my former Gatun days, was licking at the 27-foot level that morning. Under the brilliant blue sky it looked like some vast unruffled mirror—which is no figure of speech, but plain fact.

> Splendid royal palms stood up to their necks in the water . . . with their jagged noses just above the surface, gasping their last.

"Through a Forest in a Motor-boat" we might have dubbed the trip. We had soon crossed the unbroken expanse of the lake and were moving through a submerged forest. Splendid royal palms stood up to their necks in the water, corpulent, century-old giants of the jungle stood on tip-toe with their jagged noses just above the surface, gasping their last. Great mango-trees laden with fruit were descending into the flood. The lake was so mirror-like we could see the heads of drowning palm-trees and the blue sky with its wisps of snow-white feathery clouds as plainly below as above, so mirror-like the protruding stump of a palm looked like a piece of just double that length and exactly equal ends floating upright like a water thermometer, so reflective that the broken end of a branch showing above the surface appeared to be an acute angle of wood floating exactly at the angle in impossible equilibrium.

Our prisoner and crew was from "Bằhbằydos" [Barbados]—only you can't pronounce it as he did, nor make the "a" broad enough, nor show the inside of your red throat clear back to the soft palate to contrast with the glistening black skin of your carefree, grinning face. Theoretically he was being punished for assault and battery. But if this is punishment to be sentenced to cruise around on Gatun Lake I wonder crime on the Zone is so rare and unusual. This much I am sure, if I were in that particular "Bādgyän's [Barbadian's]" shoes—no, he had none; but his tracks, say—the day my time ran out I should pick a quarrel with a Jamaican and leave his countenance in such a condition that the judge could find no grounds for a reasonable doubt in the matter.

We were mounting the river Trinidad. River, yes, but we followed it only because it had kept back the jungle and left a way free of tree-tops, not because there was not water enough anywhere, in any direction, to float a boat of many times our draught. Turns so sharp we rocked in our own wake; once we passed acres upon acres of big, cod-like fish floating dead upon the water among the branches and the forest rubbish. It seems the lake in rising spread over some poisonous mineral in the soil. But life there was none, except the rampant green dying plant life in every direction to the horizon. There were not even birds, other than now and then a stray snow-white slender one of the heron species that fled majestically away across the face of the nurtureless waters as we steamed—no, gasolined down upon it. Soon after leaving Gatun we had passed a couple of jungle families on their way to market in their cayucas laden with mounds of produces—plump mangoes with a maidenly blush on either cheek, fat yellow bananas, grass-green plantains, a duck or a chicken standing tied by one leg on top of it all and gazing complacently around at the scene with the air of an experienced tourist. It was two hours later that we sighted the next human being. He was a solitary old native paddling about at the entrance to the "grass-bird region" in a huge dug-out as time-scarred as himself.

Burned Villages

It was near here that weeks before I had turned with "Admiral" B—— up a little stream now forever gone to a knoll on which sat the thatched shelter of a negro who had "taken to the bush" and refused to move even when notified that he was living on U.S. public domain. When we had knocked from the trees a box of mangoes and turkey-red *marañones* [the fruit of the cashew tree], B—— touched a match to the thatch roof and almost before we could regain the launch the shack was pouring skyward in a column of smoke. Even the squatter's old table and chair and a barrel of tumbled odds and ends entirely outside the hut—it had no walls—caught fire, and when we lost sight of the knoll only the blazing stumps of the four poles that had supported the roof remained.

> [Police] had burned whole villages in this lake territory, after the owners with legal claims had been paid condemnation damages.

B—— had burned whole villages in this lake territory, after the owners with legal claims had been paid condemnation damages. Long ago the natives had been warned to move, and the banks of the lake-to-be specified. But many of these skeptical children of nature had taken this as a vain "yanqui" boast and either refused to move until burned out or had rebuilt their hovels on land that in a few months more would also be flooded.

The Rising Lake

The rescue expedition proceeded. Once we got caught in the top-most branches of a tree, released from which we pushed on along the sinuous river that had no banks. It was not hot, even at noonday. We sweated a bit in poling a thirty-foot boat out of a tree-top, but cooled again directly we were off. My Kodak [camera] was far away at the other end of the Zone. But then, on second thought it was better for once to enjoy nature as it was without trying to carry it away. Kodaking is a species of covetousness, anyway, an attempt to bear away home with us and hoard for our own the best we come upon in our travels. Whereas here, of course, it was impossible. The greatest of artists could not have carried away a tenth of that scene, a scene so fascinating that though we had tossed into the bottom of the boat at the start a bundle of fresh New York papers—and fresh New York papers are not often scorned down on the Zone—they still lay in the bottom of the boat when the trip ended.

A village is flooded by the rise of the artificially created Gatun Lake during the Panama Canal project in 1915. (© George Shiras/National Geographic Society/ Corbis.)

At length little thatched cottages began to appear on knolls along the way, and as we chugged our way around the tree-tops upon them the inhabitants slipped quickly into some clothes that were evidently kept for just such emergencies. Then we began nearing higher land, so that the upper and then the lower branches of the forest stood out of water, then only the ends of the lower limbs dipped in the rising flood, downcast, as if they knew the sentence of death was upon them also. For though there was sunk already beneath the flood a forest greater than ten Fontainebleaus [a large forest in France], the lake was steadily rising a full two inches a day. Where it touched that morning the 27-foot level, in a few months more, says "the Colonel [chief engineer George Goethals]," it will reach the 87-foot level and spread over one hundred and sixty-four square miles of territory—and when "the Colonel" makes an assertion wise men hesitate to put their money on the other horse. Then will all this vast area with more green than in all the state of Missouri disappear forever beneath the flood and man may dive down, down into the forest and see what the world was

"Suppose a throng of unsympathetic foreigners suddenly appeared resolved to turn all the world you knew into a lake."

like in Noah's time, and fancy the sunken cities of Holland, for many a famous route, and villages older than the days of [Spanish explorer Francisco] Pizarro will be forever wiped out by the rising waters—a scene to be beheld today nowhere else, and in a few years not even here.

At last we were really in a river, an overflowed river, to be sure, where it would have been hard to find a landing-place or a bank among those tree trunks knee-deep in water. We had long since crossed the Zone line, but our badges were still valid. For it has pleased the Republic of Panama, at a whispered word from "Tio Sam [Uncle Sam]," to cede to the Z.P. [canal zone police] command over all Gatun Lake and for three miles around it, as far as ever it may spread. . . .

On a mango-shaped knoll thirty miles from Gatun that will also soon be lake bottom, we found a native shack transformed into the headquarters of a scientific expedition. . . .

By one we had Dr. O—— aboard and were waving farewell to the camp. The return, of course, was not the equal of the outward trip; even nature cannot duplicate so perfect a thing. But two raging showers gave us views of the drowning jungle under another aspect, and between them we awakened vast rolling echoes across the silent flooded world by shooting at flocks of little birds with an army rifle that would have killed an elephant.

The Plight of the Natives

It is not hard to realize why the bush native does not love the American. Put yourself in his breech-clout. Suppose a throng of unsympathetic foreigners suddenly appeared resolved to turn all the world you knew into a lake, just because that absurd outside world wanted to float steamers you never knew the use of, from somewhere you never heard of, to somewhere you did not know. Suppose a representative of that unsympathetic government came snorting down upon you one day in a wild fearful invention they called a motor-boat, as you were lolling under the thatch roof your grandfather built, and cried:

"Come on! Get out of here! We're going to burn your house and turn this country into a lake."

Flood the land which was your great-grandfather's, the spot where you used to play leap-frog under the banana trees, the jungle lane where your mother's courtship days were passed and the *ceiba* tree under which she was wedded—if matters were ever carried to that ceremonious length. What though this foreign nation gave you a bag of peculiar pieces of metal for your trouble, when you had never seen a score of such coins in your life and barely knew the use of them, being acquainted with life only as it is picked from a mango-tree? The foreigners had cried, "Take this money and go buy a farm somewhere else," and you looked around you and saw all the world you had ever really known the existence of sinking beneath the rising waters. Where would you go, think you, to buy that new farm? Even if you fled and found another unknown land high and dry, or a town, what could you do, having not the remotest idea how to live in a town with only pieces of metal to get food out of instead of the mango-tree that had stood behind the house your grandfather built ever since you were born and dropped mangoes whenever you were hungry? To say the least you would be some peeved.

It was midafternoon when the white bulk of Gatun locks rose on the horizon. Then the lake opened out, the great dam, that is rather a connecting link between two ranges of hills, spread across all the landscape, and at four I raced up the muddy steps behind the station to a telephone. Five minutes later I was hurrying away across locks and dam to the marshland beyond the Spillway to inquire who, and wherefore, had attempted to burn up the I.C.C. launch attached to dredge No. ——.

CHRONOLOGY

1850 The Clayton-Bulwer Treaty is signed by the United States and Great Britain, promising that both countries will have joint control over any future Central American canal.

1855 The United States completes a railroad across the Isthmus of Panama.

1869 US president Ulysses Grant orders surveys for a potential canal across the Isthmus of Panama.

1881 A French company begins an attempt to build a canal across Panama but fails because of the difficult terrain, inadequate equipment, and high mortality among workers from tropical diseases.

1887 A US regiment is sent to survey Nicaragua as a possible site for a canal.

1889 The French company shuts down and abandons its Panama Canal effort. A US company is asked to construct a canal through either Nicaragua or Panama. Nicaragua is chosen, and construction begins, but the effort ends in 1893 when the company goes bankrupt.

1898 During the Spanish-American War, the US battleship *Oregon* takes sixty-seven days to travel from San Francisco to the Caribbean. Though a remarkably fast passage for the era, this event demonstrates the military need for a canal to shorten the distance between the Pacific and Atlantic oceans.

1897 An Isthmian Canal Commission is appointed to study the feasibility of a US canal, and in 1899 a second Commission is appointed. Both favor a route through Nicaragua rather than Panama, as does the US public.

1901 In the Hay-Pauncefote Treaty, which nullifies the Clayton-Bulwer Treaty, the British give up any rights to a Central American canal other than for its use.

1902 A bill establishing the Nicaraguan route is introduced in the US Congress. After considerable debate it is amended and passes by a narrow margin in a form authorizing purchase of the French rights and assets in Panama.

1903 The United States negotiates a treaty with Colombia—of which Panama is then a part—for the right to build a canal across the isthmus.

March 14: The US Senate ratifies the treaty. However, Colombia fails to ratify it, demanding more money.

November 3: Panama declares its independence from Colombia, backed by US military ships sent by President Theodore Roosevelt.

November 18: The United States and Panama sign the Hay-Bunau-Varilla Treaty, giving the United States the right to construct a canal and to control a ten-mile-wide zone through which it will pass.

1904 February 23: The US Congress ratifies the Hay-Bunau-Varilla Treaty.

May 4: Americans begin preparatory work in Panama, including construction of housing, importing equipment, and undertaking sanitation measures to eradicate disease.

November 11: The first steam shovel starts work on the Culebra Cut.

1905 Yellow fever is permanently eradicated on the Isthmus of Panama, with the last case in Panama City reported on November 11.

1906 Engineers decide to build a lock canal rather than a sea-level canal. The US Congress approves this plan by a narrow margin on June 27.

1907 Construction of the canal itself begins.

1912 Congress passes a law exempting US coastwise shipping from the tolls to be assessed on vessels that use the canal. The British government protests on the ground that this violates the Hay-Pauncefote Treaty, resulting in worldwide controversy about the interpretation of that treaty.

1913 In September, US president Woodrow Wilson pushes a button in the White House that triggers the release of water into the completed canal.

1914 President Wilson asks Congress to repeal the Panama Canal tolls act to maintain the good reputation of the United States among other nations. Despite bitter political controversy, Congress complies.

1914 January 7: The first oceangoing ship travels through the Panama Canal.

August 14 (August 15 Greenwich Time): The canal officially opens with the passage of the US ship SS *Ancon.* A planned celebration is cancelled because of World War I, which began two weeks earlier.

1921 Under the leadership of President Wilson, the United States pays reparations of $25 million to Colombia for its loss of Panama.

1964 Anti-US rioting, resulting in more than twenty deaths, breaks out in the Panama Canal Zone after US students raise an American flag at Balboa High School. A recent court ruling had declared that, because of the Panamanians' wish to fly their own flag in the Zone, neither flag was to be flown. Panama breaks ties with the United States and demands a revision of the canal treaty.

1977 US president Jimmy Carter signs a treaty agreeing to give control of the canal to Panama.

1978 The US Senate votes 68–32 to turn the Panama Canal over to Panamanian control on December 31, 1999.

1979 The 1977 Panama Canal treaties become law. The United States hands over the Canal Zone, but not the canal itself, to Panama.

1999 December 31: The United States formally relinquishes complete control of the canal and its operations to Panama.

FOR FURTHER READING

Books

Willis J. Abbott, *Panama and the Canal: The Story of Its Achievement, Its Problems and Its Prospects.* New York: Dodd Mead, 1914.

Ralph Emmet Avery, *America's Triumph at Panama.* Chicago: L.W. Walter Co., 1913.

John Barrett, *Panama Canal: What It Is, What It Means.* Washington, DC: Pan American Union, 1913.

Joseph Bucklin Bishop, *The Panama Gateway.* New York: Scribner's, 1913.

Joseph Bucklin Bishop, *Theodore Roosevelt and His Time Shown in His Own Letters*, vol. 1. New York: Scribner's, 1920.

Philippe Bunau-Varilla, *Panama: The Creation, Destruction and Resurrection*, New York: McBride, Nast & Co., 1914.

David Newton E. Campbell, *Searchlight on the Panama Canal, or America's Greatest Enterprise.* Baltimore: Baltimore Book Co., 1909.

Julie Green, *The Canal Builders: Making America's Empire at the Panama Canal.* New York: Penguin, 2009.

Ulrich Keller, *The Building of the Panama Canal in Historic Photographs.* Mineola, NY: Dover, 1984.

Walter LaFeber, *The Panama Canal: The Crisis in Historical Perspective.* New York: Oxford University Press, 1989.

David McCullough, *The Path Between the Seas: The Creation of the Panama Canal, 1870–1914.* New York: Simon & Schuster, 1977.

Duncan E. McKinlay, *The Panama Canal.* San Francisco: Whitaker & Ray-Wiggin, 1912.

Noel Maurer and Carlos Yu, *The Big Ditch: How America Took, Built, Ran, and Ultimately Gave Away the Panama Canal.* Princeton, NJ: Princeton University Press, 2011.

Hugh Gordon Miller and Joseph C. Freehoff, *The Panama Canal Tolls Controversy.* Boston: Chapple Publishing Co., 1914.

J. Saxon Mills, *The Panama Canal: A History and Description of the Enterprise.* London: Thomas Nelson and Sons, 1913, and Echo Library, 2011.

Alexander Missal, *Seaway to the Future: American Social Visions and the Construction of the Panama Canal.* Madison: University of Wisconsin Press, 2008.

Matthew Parker, *Panama Fever: The Epic Story of the Building of the Panama Canal.* New York: Doubleday, 2007.

William R. Scott, *The Americans in Panama.* New York: Statler, 1912.

Jeremy Sherman Snapp, *Destiny by Design: The Construction of the Panama Canal.* Lopez Island, WA: Pacific Heritage Press, 2000.

Rose Van Hardeveld, *Make the Dirt Fly!* Hollywood, CA: Pan Press, 1956.

Hugh C. Weir, *The Conquest of the Isthmus: The Men Who Are Building the Panama Canal—Their Daily Lives.* New York: Putnam, 1909.

Periodicals and Internet Sources

Jimmy Carter, Address to the Nation on the Panama Canal Treaties, February 1, 1978. www.presidency.ucsb.edu.

Jimmy Carter, "Why I Gave Away the Panama Canal," *Time,* October 18, 1982. www.time.com.

Centers for Disease Control and Prevention, "Malaria: The Panama Canal," February 8, 2010. www.cdc.gov.

Adam Clymer, "A Movement, a Plan, a Canal," *New York Times,* March 15, 2008. www.nytimes.com.

Neal Creighton, "The Panama Canal Treaty—An Update," *National Strategy Forum Review*, January 6, 2011. http://nationalstrategy.com.dnnmax.com.

Henry Fountain, "Panama Adding a Wider Shortcut for Shipping," *New York Times*, August 16, 2011. www.nytimes.com.

Alexandra H. Freeman, "The Mosquito of High Crimes: The Campaign Against Yellow Fever During the American Construction of the Panama Canal, 1904–1905," *Historia Medicinae,* August 17, 2011. www.medicinae.org.

Sarah Jane Gilbert, "Troubled History, Astounding Turnaround," Harvard Business School Working Knowledge, December 20, 2010. http://hbswk.hbs.edu.

GlobalSecurity.org, "Panama Canal," www.globalsecurity.org.

Serge F. Kovaleski, "Ceremony Marks US Handover of Panama Canal, *Washington Post*, December 15, 1999. www.latinamericanstudies.org.

Marjory Burns Shanard, "Panama Canal Treaties," Panama Canal Museum, 1977. http://panamacanalmuseum.org.

Jay Stuart Snelson, "The Ideological Immune System: Resistance to New Ideas in Science," eSkeptic, January 26, 2011. www.skeptic.com.

"Theodore Roosevelt and the Panama Canal," *American Experience: TR*, PBS, 1996. www.pbs.org/wgbh/americanexperience.

Edwin Warner, "That Troublesome Panama Canal Treaty," *Time*, October 31, 1977. www.time.com.

James R. Whelan, "Reversion? Hogwash," *Human Events*, November 12, 1999.

Websites

American Experience: Panama Canal (www.pbs.org/wgbh/americanexperience/films/panama). A ninety-minute PBS program that can be watched online, with a transcript as well as related feature articles.

Canal Museum (www.canalmuseum.com). A website "dedicated to all men and women who have and will work at or for the Panama Canal and made the heroic dream of over 400 years come true." It contains both historical and contemporary photos and stories.

Canal Zone Brats (www.czbrats.com). A website maintained by Americans who grew up in the Canal Zone. It contains material about the history of Panama and the canal along with photos, articles, stories, current material, and information about high school class reunions.

Canal Zone Images (www.czimages.com). Site of the Panama Canal Zone Cyber Museum, dedicated to preserving the heritage of "Zonians," with pictures and memorabilia from various eras and links to other Panama Canal sites.

Global Perspectives: Panama Canal (www.cotf.edu/earthinfo/camerica/panama/PCtopic1.html). A site containing detailed information about the canal and United States-Panamanian relations since 1914.

Panama Canal Authority (www.pancanal.com/eng/acp). A site maintained by the Panamanian organization now in charge of the canal, containing extensive information about its current status, as well as many pictures and historical information.

Panama Canal Museum (http://panamacanalmuseum.org). A museum founded to preserve the history of Americans in Panama with a focus on the Panama Canal. Its website contains information about its collection, events, personal stories, and links to related material.

Panama Canal Society (www.pancanalsociety.org). An organization of individuals who formerly worked on the canal, family members who resided there, and their descendants. It contains many personal stories of life in the Canal Zone from the 1920s through the 1960s.

INDEX

L

M

N

O

P

R

S

T

U